Plasma Isoenzymes: The Current Status

Advances in Clinical Enzymology

Vol. 3

Series Editors
D.W. Moss, London, *E. Schmidt* and *F.W. Schmidt,* Hannover

KARGER

Basel · München · Paris · London · New York · New Delhi · Singapore · Tokyo · Sydney

Proceedings of the 3rd International Symposium of the Belgian Society of Clinical Chemistry, Brugge, October 24–26, 1985

Plasma Isoenzymes: The Current Status

Volume Editors
V. Blaton and *A. Van Steirteghem,* Brugge

43 figures and 38 tables, 1986

KARGER

Basel · München · Paris · London · New York · New Delhi · Singapore · Tokyo · Sydney

Advances in Clinical Enzymology

National Library of Medicine, Cataloging in Publication
Société belge de chimie clinique. International Symposium (3rd: 1985: Brugge, Belgium)
Plasma isoenzymes: the current status: proceedings of the 3rd International Symposium of the Belgian Society of Clinical Chemistry, Brugge, October 24–26 1985/volume editors, V. Blaton and A. Van Steirteghem. – Basel; New York: Karger, 1986. – (Advances in clinical enzymology; vol. 3)
Includes index.
1. Isoenzymes – blood – congresses I. Blaton, V. II. Steirteghem, A. Van III. Title IV. Series
W1 AD54J v.3
QU 135 S6775 1985p
ISBN 3-8055-4321-2

Drug Dosage

The authors and the publisher have exerted every effort to ensure that drug selection and dosage set forth in this text are in accord with current recommendations and practice at the time of publication. However, in view of ongoing research, changes in government regulations, and the constant flow of information relating to drug therapy and drug reactions, the reader is urged to check the package insert for each drug for any change in indications and dosage and for added warnings and precautions. This is particularly important when the recommended agent is a new and/or infrequently employed drug.

Printed in Switzerland by Thür AG Offsetdruck, Pratteln
ISBN 3-8055-4321-2

Contents

Preface

Ever since the first clear statement of the isoenzyme concept as a major generalization in biochemistry, now rather more than 25 years ago, the possibility of exploiting isoenzymes and other multiple forms of enzymes in the diagnosis and understanding of disease has been of major interest to clinical enzymologists. Indeed, much of the impetus of isoenzyme research has come from clinical laboratories. It was in this tradition that the Belgian Society for Clinical Chemistry devoted a major part of its Third International Symposium, held in Brugge, October 24–26, 1985, to a review of the current status of isoenzyme applications and research in clinical and laboratory medicine.

A number of isoenzyme systems (using this term in an operational sense to include multiple forms which arise from postgenetic causes as well as genetically determined isoenzymes) are now well established in diagnostic medicine; in particular, alkaline phosphatase and creatine kinase. These systems are well represented. In addition, isoenzymes such as those of amylase which are beginning to make their impact in the clinical laboratory due to advances in methodology are discussed, as are the intriguing variants of γ-glutamyltransferase which seem to arise entirely by modifications of the enzyme encoded by a single gene. Other enzyme systems which are less frequently discussed in the context of isoenzymes are not neglected, such as the glycolytic enzymes, and enzymes involved in inflammation. Finally, since all clinical scientists hope to progress from the diagnosis and study of disease to its treatment, we include a review of lysosomal enzyme replacement therapy as one example of how this progression is already being achieved.

We are confident that all clinical scientists who are concerned, even indirectly, with clinical enzymology will find much to interest them in these up-to-date and authoritative reviews.

V. Blaton
A. Van Steirteghem
D.W. Moss (series editor)

Adv. clin. Enzymol., vol. 3, pp. 1–11 (Karger, Basel 1986)

Alkaline Phosphatase Isoenzymes

The Current Status

D.W. Moss

Royal Postgraduate Medical School, Hammersmith Hospital, London, UK

The human alkaline phosphatases (EC 3.1.3.1) exist in numerous multiple forms, some of which are true isoenzymes encoded by distinct structural genes with localized expressions, while others result from co-translational or post-translational modifications (table I). In particular, post-translational modifications of the product of a widely-expressed, 'tissue-unspecific' structural gene give rise to tissue-specific forms. Both the true isoenzymes of alkaline phosphatase and the post-translationally modified forms (usually referred to collectively as isoenzymes) are of clinical importance: in sensitive and organ-specific diagnosis, in tumour detection and localization, and in targetted tumour therapy. Furthermore, the expression of alkaline phosphatase in cell lines in vitro provides a model system in which the factors affecting specific isoenzyme expression can be studied. Much of the impetus of alkaline phosphatase isoenzyme research has therefore come from clinical observation, and the unravelling of the complex patterns of enzyme forms present in tissues and serum has gone hand in hand with clinical exploitation of established tissue-specific differences, even while the underlying nature of these differences has remained unknown.

Properties of Multiple Forms of Alkaline Phosphatase

All the alkaline phosphatases are of wide substrate specificity, catalysing the removal of the terminal phosphate groups (i.e. those with two unesterified hydroxyl groups) from a wide range of orthophosphate and pyro-

Table I. Probable genetic origins and expression of multiple forms of human alkaline phosphatase: intra-tissue heterogeneity arising from variations in carbohydrate moieties and aggregation phenomena also exists

Isoenzymes encoded by independent genetic loci	Expression
'Tissue-unspecific'	expressed widely in, for example, bone, liver, kidney, as variants with tissue-specific properties due to differences in glycosylation of presumed identical protein
Adult intestinal	localized to small-intestinal brush border; traces in other tissues, e.g. kidney
Fetal intestinal (status as product of a separate structural gene not yet completely proved, but probable)	expressed in fetal intestine up to about 30 weeks of gestation
Placental	highly localized expression in form of numerous allelozymes in term placenta; traces in other tissues, e.g. lung, cervix
Testicular (status as product of a separate gene or as a variant of placental isoenzyme still undecided)	similar to placental isoenzyme except with some monoclonal antibodies and in inhibition by *L*-leucine; expressed in trace amounts in testis, thymus, and other tissues

phosphate substrates, including the cleavage of inorganic pyrophosphate [7, 22, 25]. Although the reaction is usually one of hydrolysis, phosphate transfer to the hydroxyl group of a suitable acceptor is also catalysed. However, as is generally the case with isoenzymes encoded by multiple gene loci, there are quantitative and qualitative differences in catalytic properties between the true isoenzymes of alkaline phosphatase. For example, the intestinal and placental alkaline phosphatases are similar in their relative rates of action on alternative substrates, but differ in this respect from the phosphatases of such tissues as bone and liver, which appear to be encoded by a single, tissue-unspecific gene. The various products of the tissue-unspecific gene resemble each other in their catalytic properties. The ratio of rates of hydrolysis of two substrates has been used for the non-separative determination of intestinal phosphatase in serum [20]. Placental and intestinal phosphatases are again similar, but distinct from other phosphatases, in their reactions with stereo-specific, uncompetitive inhibitors such as *L*-phenylalanine or certain peptides. *L*-Phenylalanine markedly inhibits these two isoenzymes, but has less effect on the tissue-unspecific phosphatases [9]. A converse pattern of inhibition is seen with *L*-homoarginine and levamisole [35]. Uncompetitive inhibition by amino acids has been widely used in characterization and non-separative determination of intestinal and placental alkaline phosphatases, and in purification by affinity chromatography.

In spite of their similarities with respect to alternative substrates and uncompetitive inhibitors, the intestinal and placental isoenzymes are not identical in their ligand-binding properties, as is shown by their reactions with the dye Reactive Yellow 13. This compound in solution inhibits the phosphate-transfer reaction catalysed by the alkaline phosphatases; i.e. it converts the reaction from one of hydrolysis plus phosphate transfer, to one of hydrolysis alone [43]. However, the inhibition of intestinal phosphatase is competitive with respect to the phosphate acceptor (e.g. diethanolamine), but uncompetitive in the case of the other alkaline phosphatases, including the placental isoenzyme [43]. Thus, a binding-site for Reactive Yellow 13 exists in intestinal phosphatase in the absence of the substrate or phosphate acceptor so that, when immobilized on a matrix such as Sepharose 4B, Reactive Yellow 13 acts as a specific affinity adsorbent for intestinal phosphatase [42]. The isoenzyme can be eluted quantitatively by a concentrated solution of diethanolamine. Adult and fetal intestinal phosphatases appear to have closely similar or even identical ligand binding sites.

As is the case for other isoenzyme systems, thc most widely-used technique for identifying alkaline phosphatases is zone electrophoresis, particularly in the analysis of serum specimens. The enzymes from liver, bone, kidney and small intestine display characteristic electrophoretic patterns, including the micro-heterogeneity typical of glycoproteins. The electrophoretic mobility of placental phosphatase depends on the allelozyme(s) present, and electrophoresis has been essential in the discovery of the polymorphism of this enzyme [6].

The difference in mobility is one of the few characteristics which distinguish between liver and bone phosphatases; consequently, in view of their diagnostic importance, much effort has been devoted to improving their electrophoretic resolution. Although complete digestion with neuraminidase abolishes the difference in mobility between liver and bone phosphatases [18, 31], minimal neuraminidase digestion accentuates the difference between them [4, 11, 23] and can be used to improve resolution in polyacrylamide-gel electrophoresis [23]. Affinity electrophoresis in the presence of wheat-germ lectin also increases the difference in mobility between bone and liver phosphatases on cellulose acetate electrophoresis [30].

Placental alkaline phosphatase is remarkable among human enzymes in resisting denaturation at 65 °C. There are small but consistent differences in stability to heat [24] or urea [1] between the tissue-unspecific alkaline phosphatases. The most marked of these are between the bone and liver phosphatases, which have average half-inactivation times in sera at 56 °C of 112 and 456 s, respectively [41]. As with other properties of these enzymes, there is significant sample-to-sample variation: the respective standard deviations of the half-inactivation times are 25 and 90 s [41]. For this reason, multi-point methods are preferable to determination of residual activity after a single period of heating when quantitatively determining liver and bone phosphatases in serum by heat inactivation. With suitable experimental precautions, multi-point heat-inactivation methods are capable of excellent reproducibility and can be regarded as reference methods against which other procedures can be judged [27]. Since other evidence strongly supports the hypothesis that the tissue-unspecific alkaline phosphatases have a common protein core, their differences in stability, like those in charge-dependent properties, must originate from differences in their carbohydrate moieties. This is supported by recent observations that the heat stability of liver phosphatase is reduced by digestion with *Trichomonas foetalis* glycosidase, so that it approaches that of the less-stable bone phosphatase [26].

As is the case for true isoenzymes in general, those of alkaline phosphatase (i.e. the placental, intestinal and tissue-unspecific forms) are antigenically distinct. Polyclonal antisera distinguish between these three forms, although there is typically some cross-reaction between the placental and intestinal antisera which can be eliminated by absorbing-out [17]. There is little or no cross-reaction between these isoenzymes and the phosphatases of the tissue-unspecific group, but polyclonal antisera fail to differentiate at all between the tissue-unspecific phosphatases.

Monoclonal antibodies are able to distinguish between the allelozymes of placental alkaline phosphatase [32]. It is also largely through the availability of these reagents that a ‘placental-like’ alkaline phosphatase, resembling true placental phosphatase in all but a few properties, has been recognized in trace amounts in normal testis and certain other tissues, and as the product of some tumours [34]. The development of monoclonal antibodies specific for bone or liver alkaline phosphatases has proved to be more difficult, as would be expected from the presumed identity of their protein components. However, reports are appearing that even this problem is beginning to yield to the exquisite sensitivity of monoclonal antibodies as structural probes, although the discrimination obtained is by no means complete [16]. Furthermore, monoclonal antibodies able to distinguish between fetal and adult intestinal alkaline phosphatases have been obtained [37].

The earliest inferences regarding the possible structural differences between multiple forms of alkaline phosphatase were based on ligand-binding and selective modification studies of the type already mentioned. The introduction of monoclonal antibodies has increased the number of recognizable differences between alkaline phosphatase molecules, but without delineating the structures of the several epitopes with which the antibodies react. Partial structural analysis of alkaline phosphatase proteins has included determination of N-terminal amino acid sequences, and peptide mapping of isoenzymes radiolabelled either unspecifically, or specifically at the active centre with radioactive orthophosphate, to facilitate the identification of peptides derived from the isoenzymes [18, 31, 39]. Studies of this kind have led to the identification of at least three distinct protein classes, represented by the placental, intestinal and tissue-unspecific alkaline phosphatases, presumed to be encoded by at least three structural genes. The existence of a separate gene which encodes the 'placental phosphatase-like' enzyme remains uncertain. However, partial structural comparisons between adult and fetal intestinal alkaline phosphatases, based on cyanogen-bromide peptide mapping [36] and maps of tryptic digests of radio-iodinated enzymes [28], suggest that these phosphatases are products of distinct structural genes.

The advances in the determination of protein primary structures brought about by gene cloning and sequencing techniques are already beginning to make their impact on alkaline phosphatase research: the gene encoding one of the allelozymes of human placental alkaline phosphatase has been successfully cloned and the primary structure of the protein deduced [19]. The structure of the carbohydrate side chains has also been determined. It therefore seems probable that full structural analyses of the major alkaline phosphatase variants will follow in the foreseeable future.

Clinical Applications of Alkaline Phosphatase Characterization

Since a raised serum alkaline phosphatase activity is characteristic of two main groups of diseases, bone diseases and hepatobiliary diseases, the most common reason for requesting alkaline phosphatase isoenzyme analysis in diagnosis is the need to distinguish between bone and liver as alternative, or co-existing, sources of the raised serum activity. More than half of all requests probably arise in this way, and can usually be answered by qualitative techniques, such as the visual inspection of zymograms. However, quantitative methods are needed fully to exploit the potential increase in sensitivity, as well as organ specificity, offered by isoenzyme analysis. This is especially true when searching for minimum pathological changes, or in following the progress of disease, for example in patients with metastatic cancer [21]. Indeed, the use of specific, quantitative determination of bone alkaline phosphatase in the diagnosis and treatment of cancers metastatic to bone is possibly the most important clinical application of alkaline phosphatase isoenzyme analysis, since even those cancers which give rise to predominantly osteoclastic metastases usually stimulate an osteoblastic response, detectable by an increase in bone phosphatase in serum provided that suitable methods of analysis are used. Furthermore, isoenzyme determinations are free from the risks associated with repeated applications of imaging techniques. Quantitative analysis by three-point selective heat inactivation has been found to meet this need admirably in our laboratory during the past ten years [27, 40].

The occurrence in serum of a high molecular weight, highly negatively-charged form of alkaline phosphatase (the 'fast-liver' fraction) has been advocated as an index of biliary obstruction, particularly of the localized intrahepatic obstruction caused by malignant infiltration of the liver. However, this fraction seems to be neither more sensitive nor more specific in this respect than the main liver fraction itself. The high molecular weight alkaline phosphatase in serum is heterogeneous, part corresponding to enzyme bound to membrane fragments (koinozymes) [5] and part probably resulting from interaction between the enzyme and lipoproteins [3]. Analogous high molecular weight forms are usually detectable in tissue extracts.

The levels of bone and liver phosphatases in serum are governed by their rates of entry into the circulation from the cells in which they are produced. In contrast, the level of intestinal alkaline phosphatase, which reaches the circulation in large amounts via the thoracic lymph [13], is determined largely by its rate of removal. Usually this is extremely rapid, so

Table II. Tumour-associated alkaline phosphatases (ALP) and their putative normal counterparts

Tumour product	Characteristic properties	Normal counterpart
Placental ALP (Regan isoenzyme)	heat-stable; *L*-phenylalanine sensitive; antigenically distinct	term-placental ALP
Placental ALP-like (Nagao isoenzyme)	heat-stable; *L*-phenylalanine and *L*-leucine sensitive; differs from P1-ALP with m-Abs	placental ALP-like enzyme from testis, etc.
Intestinal ALP-like (Kasahara isoenzyme, hepatoma-associated Regan variant)	heat-labile; *L*-phenylalanine and *L*-leucine sensitive; cross-reacts with anti-intestinal ALP antisera; sialylated	fetal intestinal ALP
Tissue-unspecific (non-Regan)	heat-labile; *L*-homoarginine sensitive; various degrees of sialylation	bone, liver, kidney, etc. ALPs

that only small activities of the intestinal isoenzyme are detectable in serum, even in subjects of blood groups B or O who are secretor positive [15], in whom clearance of the isoenzyme seems to be somewhat delayed. There is, therefore, a poor correlation between serum intestinal alkaline phosphatase and disease, although increased levels are encountered, e.g. in chronic liver disease, especially in patients of the appropriate blood groups. Recently, a low level of fetal intestinal alkaline phosphatase in amniotic fluid at 16–18 weeks of gestation has been shown to be a reliable indication of cystic fibrosis in the fetus [2].

Inappropriate expression in cancer tissues of alkaline phosphatases which correspond more or less closely to normal forms of the enzyme is now a well-recognized phenomenon (table II). Placental alkaline phosphatases ('Regan isoenzymes') are present in sera from 5–15% of patients with cancers of all types, with higher incidences in, for example, genital tumours, and in tumour tissues [33]. A significant advance in the detection of placental and placental-like alkaline phosphatases in cancer has resulted from the availability of monoclonal antibodies. Like their normal counterparts, tumour-associated placental phosphatases exist as molecular variants, and the use of a panel of monoclonal antibodies is more effective than a single antibody in detecting the isoenzymes in cancer patients. Monoclonal antibodies have also facilitated the recognition of a 'placental-like' isoenzyme in trace amounts in normal tissues such as testis, and in increased amounts ('Nagao isoenzymes') in sera and tissues of patients with certain cancers, particularly dysgerminoma of the ovary and seminoma in which the isoenzyme is a particularly valuable tumour marker [8]. Monoclonal antibodies directed against placental phosphatase and placental-like phosphatase are opening up new possibilities of specific tumour localization and of targetted cancer therapy with drugs or radioisotopes.

Other alkaline phosphatase tumour markers include the 'Kasahara isoenzymes', which are associated particularly, but not exclusively, with hepatoma [10, 38] and which appear to correspond in properties to normal fetal intestinal phosphatase.

Current Problems

The isoenzyme status of the placental-like alkaline phosphatase remains to be defined: its separate identity so far rests on its reaction with monoclonal antibodies, so that independent evidence of its individuality is eagerly awaited. In the case of fetal intestinal phosphatase, although partial structural analysis as well as distinctive reaction with monoclonal antibodies provide strong evidence of a unique protein structure probably encoded by a unique structural gene, conclusive proof is also still required.

Among the tissue-unspecific alkaline phosphatases, the mechanisms of the postulated tissue-specific differences in glycosylation remain to be elucidated. Furthermore, although it is now clear that bile stasis-induced synthesis of hepatic alkaline phosphatase is responsible for the observed rapid increase in serum alkaline phosphatase in obstructive jaundice [12, 14], the nature of the inducer and its mode of action remain unknown.

References

1 Birkett, D.J.; Conyers, R.A.J.; Neale, F.C.; Posen, S.; Brudenell-Woods, J.: Action of urea on human alkaline phosphatases with a description of some automated techniques for the study of enzyme kinetics. Archs Biochem. Biophys. *124:* 470–479 (1967).

2 Brock, D.J.H.; Barron, L.; Bedgood, D.; Hayward, C.: Prospective prenatal diagnosis of cystic fibrosis. Lancet *i:* 1175–1178 (1985).

3 Brocklehurst, D.; Wilde, C.E.; Doar, J.W.H.: The incidence and likely origins of serum particulate alkaline phosphatase and lipoprotein-X in liver disease. Clinica chim. Acta *88:* 509–515 (1978).

4 Bücher, T.; Dünnwald, M.; Fuhrer, F.; Linke, I.; Keller, H.; Keller, B.: Quantitative evaluation of serum alkaline phosphatase isoenzyme patterns. Abstracts, Joint Meet., Belgian, Dutch, German and British Soc. Clinical Chemistry (1983).

5 De Broe, M.E.; Borgers, M.; Wieme, R.J.: The separation and characterization of liver plasma membrane fragments circulating in the blood of patients with cholestasis. Clinica chim. Acta *59:* 369–372 (1975).

6 Donald, L.J.; Robson, E.B.: Rare variants of placental alkaline phosphatase. Ann. hum. Genet. *37:* 303–313 (1974).

7 Eaton, R.H.; Moss, D.W.: Organic pyrophosphates as substrates for human alkaline phosphatases. Biochem. J. *105:* 1307–1312 (1967).

8 Epenetos, A.A.; Munro, A.J.; Tucker, D.F.; Gregory, W.; Duncan, W.; MacDougall, R.H.; Faux, M.; Travers, P.; Bodmer, W.F.: Monoclonal antibody assay of serum placental alkaline phosphatase in the monitoring of testicular tumours. Br. J. Cancer *51:* 641–644 (1985).

9 Fishman, W.H.; Sie, H.G.: Organ-specific inhibition of human alkaline phosphatase isoenzymes of liver, bone, intestine and placenta: *L*-phenylalanine, *L*-tryptophan and *L*-homoarginine. Enzymologia *41:* 141–167 (1971).

10 Higashino, K.; Otani, R.; Kudo, S.; Yamamura, Y.: A fetal intestinal-type alkaline phosphatase in hepatocellular carcinoma tissue. Clin. Chem. *23:* 1615–1623 (1977).

11 Hill, P.G.; Sammons, H.G.: An interpretation of the elevation of serum alkaline phosphatase in disease. J. clin. Path. *20:* 654–659 (1967).

12 Kaplan, M.M.; Righetti, A.: Induction of rat liver alkaline phosphatase: the mechanism of the serum elevation in bile duct obstruction. J. clin. Invest. *49:* 508–516 (1970).

13 Keiding, N.R.: The alkaline phosphatase fractions of human lymph. Clin. Sci. *26:* 291–297 (1964).

14 Kryszewski, A.J.; Neale, G.; Whitfield, J.B.; Moss, D.W.: Enzyme changes in experimental biliary obstruction. Clinica chim. Acta *47:* 175–182 (1973).

15 Langman, M.J.S.; Leuthold, E.; Robson, E.B.; Harris, J.; Luffman, J.E.; Harris, H.: Influence of diet on the intestinal component of serum alkaline phosphatase in people of different ABO blood groups and secretor status. Nature, Lond. *212:* 41–43 (1966).

16 Lawson, G.M.; Katzmann, J.A.; Kimlinger, T.K.; O'Brien, J.F.: Isolation and preliminary characterization of a monoclonal antibody that interacts preferentially with the liver isoenzyme of human alkaline phosphatase. Clin. Chem. *31:* 381–385 (1985).

17 Lehmann, F.G.: Immunological relationship between human placental and intestinal alkaline phosphatase. Clinica chim. Acta *65:* 257–261 (1975).

18 McKenna, M.J.; Hamilton, T.A.; Sussman, H.H.: Comparison of human alkaline phosphatase isoenzymes. Structural evidence for three protein classes. Biochem. J. *181:* 67–73 (1979).

19 Millan, J.L.; Fishman, W.H.: The primary structure of human placental alkaline phosphatase (in preparation).
20 Moss, D.W.: Determination of intestinal alkaline phosphatase in serum by substrate-specificity measurements. Clinica chim. Acta *35:* 413–419 (1971).
21 Moss, D.W.: Alkaline phosphatase isoenzymes. Clin Chem. *28:* 2007–2016 (1982).
22 Moss, D.W.; Eaton, R.H.; Smith, J.K.; Whitby, L.G.: Association of inorganic-pyrophosphatase activity with human alkaline-phosphatase preparations. Biochem. J. *102:* 53–57 (1967).
23 Moss, D.W.; Edwards, R.K.: Improved electophoretic resolution of bone and liver alkaline phosphatases resulting from partial digestion with neuraminidase. Clinica chim. Acta *143:* 177–182 (1984).
24 Moss, D.W.; King, E.J.: Properties of alkaline phosphatase fractions separated by starch-gel electrophoresis. Biochem. J. *84:* 192–195 (1962).
25 Moss, D.W.; Walli, A.K.: Intermediates in the hydrolysis of ATP by human alkaline phosphatase. Biochim. biophys. Acta *191:* 476–477 (1969).
26 Moss, D.W.; Whitaker, K.B.: Modification of alkaline phosphatases by treatment with glycosidases. Enzyme (in press, 1986).
27 Moss, D.W.; Whitby, L.G.: A simplified heat-inactivation method for investigating alkaline phosphatase isoenzymes in serum. Clinica chim. Acta *61:* 63–71 (1975).
28 Mueller, H.D.; Leung, H.; Stinson, R.A.: Different genes code for alkaline phosphatases from human fetal and adult intestine. Biochem. biophys. Res. Commun. *126:* 427–433 (1985).
29 Nakayama, T.; Yoshida, M.; Kitamura, M.: *L*-Leucine sensitive, heat-stable alkaline-phosphatase isoenzyme detected in a patient with pleuritis carcinomatosa. Clinica chim. Acta *30:* 546–548 (1970).
30 Rosalki, S.B.; Foo, A.Y.: Two new methods for separating and quantifying bone and liver alkaline phosphatase isoenzymes in plasma. Clin. Chem. *30:* 1182–1186 (1984).
31 Seargeant, L.E.; Stinson, R.A.: Evidence that three structural genes code for human alkaline phosphatases. Nature, Lond. *281:* 152–154 (1979).
32 Slaughter, C.A.; Coseo, M.C.; Cancro, M.P.; Harris, H.: Detection of enzyme polymorphism by using monoclonal antibodies. Proc. natn. Acad. Sci. USA *78:* 1124–1128 (1981).
33 Stolbach, L.L.; Krant, M.J.; Fishman, W.H.: Ectopic production of an alkaline phosphatase isoenzyme in patients with cancer. New Engl. J. Med. *281:* 757–762 (1969).
34 Travers, P.; Bodmer, W.: Preparation and characterization of monoclonal antibodies against placental alkaline phosphatase and other human trophoblast-associated determinants. Int. J. Cancer *33:* 633–641 (1984).
35 Van Belle, H.: Alkaline phosphatase. I. Kinetics and inhibition by levamisole of purified isoenzymes from humans. Clin. Chem. *22:* 972–976 (1976).
36 Vockley, J.; D'Souza, M.P.; Foster, C.J.; Harris, H.: Structural analysis of human adult and fetal alkaline phosphatases by cyanogen bromide peptide mapping. Proc. natn. Acad. Sci. USA *81:* 6120–6123 (1984).
37 Vockley, J.; Meyer, L.J.; Harris, H.: Differentiation of human adult and fetal intestinal alkaline phosphatases with monoclonal antibodies. Am. J. hum. Genet. *36:* 987–1000 (1984).

38 Warnock, M.L.; Reisman, R.: Variant alkaline phosphatase in human hepatocellular cancers. Clinica chim. Acta *24:* 5–11 (1969).
39 Whitaker, K.B.; Moss, D.W.: Comparison of radioactive peptides obtained from specifically labelled renal and placental alkaline phosphatases. Biochem. J. *183:* 189–192 (1979).
40 Whitaker, K.B.; Whitby, L.G.; Moss, D.W.: Activities of bone and liver alkaline phosphatases in serum in health and disease. Clinica chim. Acta *80:* 209–220 (1977).
41 Whitby, L.G.; Moss, D.W.: Analysis of heat inactivation curves of alkaline phosphatase isoenzymes in serum. Clinica chim Acta *59:* 361–367 (1975).
42 Williams, D.G.; Byfield, P.G.H.; Moss, D.W.: Affinity chromatography of human intestinal alkaline phosphatase. Enzyme *28:* 28–32 (1982).
43 Williams, D.G.; Byfield, P.G.H.; Moss, D.W.: Inhibition of human alkaline phosphatase isoenzymes by the affinity reagent Reactive Yellow 13. Enzyme *33:* 70–74 (1985).

D.W. Moss, DSc, University of London, Royal Postgraduate Medical School,
Hammersmith Hospital, Ducane Road, London W12 OHS (UK)

Adv. clin. Enzymol., vol. 3, pp. 12–18 (Karger, Basel 1986)

Some Clinical Applications of Plasma Alkaline Phosphatase Isoenzyme Determinations

S.B. Rosalki

Department of Chemical Pathology and Human Metabolism, Royal Free Hospital, Pond Street, London, UK

Alkaline phosphatase (ALP; EC 3.1.3.1) occurs in tissue-specific multiple molecular (isoenzyme) forms which are released into the plasma, where they can be separated by electrophoresis and demonstrated by enzyme-staining techniques. This paper considers information that can be derived from electrophoretic separation and quantitation of the various isoenzyme fractions, illustrating this with studies from the author's laboratory.

Liver and Bone ALP

Liver and bone disease are the major causes of plasma ALP elevation. The liver and bone fractions, both of which migrate in the alpha$_2$-globulin position, are poorly separated by conventional electrophoresis. Because of this, it has hitherto been difficult for the routine laboratory to measure simply and reliably the activity of these isoenzymes in plasma. A new electrophoretic procedure was developed which permits this [1], based on the observation that wheat germ lectin preferentially binds alkaline phosphatase of bone origin, and can separate it from the liver isoenzyme. In this procedure, serum is separated on cellulose acetate membrane with wheat germ lectin added to the membrane buffer. After electrophoresis, alkaline phosphatase activity is demonstrated by staining with a chromogenic substrate and isoenzymes are quantified by densitometric scanning.

Table I. ALP isoenzymes in 135 patients investigated for metastases following breast carcinoma

		Percentage of abnormal results		
		total ALP	bone ALP	liver ALP
1	Bone metastases only (n = 44)	23	41	5
2	Bone + liver metastases (n = 22)	55	41	41
3	No bone, no liver metastases (n = 69)	3	3	6

This new procedure was used for the detection of liver and bone metastases in 135 patients following breast carcinoma, and serially studied clinically and by radiology, computerised tomography, scintigraphy and ultrasound, as appropriate [Mayne, Thakkar, Rosalki, Foo, Parbhoo, unpubl.]. When bone secondaries only were present, the activity of the bone isoenzyme was increased in 41% of patients (table I). This sensitivity, though modest, was almost twice that of total alkaline phosphatase. Liver ALP activity was only rarely raised in the absence of liver secondaries, but when these were present, it was increased in 41% of patients. All patients with liver metastases additionally had bone secondaries, and, presumably because of this, a slightly higher frequency of total ALP elevation was observed. However, isoenzyme examination added specificity.

Serial studies increased the sensitivity of isoenzyme measurement for metastases detection, sometimes with lead-in times of up to three to six months before diagnostic confirmation, particularly with liver metastases. The ability of the isoenzyme procedure to measure liver and bone ALP independently was especially valuable for serial monitoring.

Biliary ALP

So-called 'biliary' alkaline phosphatase is a high molecular mass ALP migrating in the alpha$_1$-globulin position by conventional electrophoresis on cellulose acetate. It is not normally detectable in plasma by electrophoresis, but may be found in patients with cholestasis or hepatic neoplasia. It is thought to derive from membrane fragments of liver and biliary tract cells or the aggregation in bile or plasma of liver ALP with lipid and protein [2].

Table II. Biliary ALP by electrophoresis and GGT in 145 breast cancer patients investigated for liver metastases

	Metastases present, % (n = 44)	Metastases absent, % (n = 101)
Total ALP ↑	43	7
Biliary ALP	57	14
GGT ↑	81	32

Since biliary phosphatase is known to be raised in malignant involvement of the liver [3], the diagnostic value of its determination for the detection of liver metastases following breast cancer was examined using an electrophoretic method validated by chromatography [4]. 145 consecutive anicteric patients without hepatomegaly were followed serially [Karmen, Mayne, Rosalki, Foo, Parbhoo, unpubl.]. Biliary ALP was increased in 57% of patients with liver metastases (table II). It was more sensitive than total or liver alkaline phosphatase, and though less sensitive than gamma glutamyltransferase (GGT; EC 2.3.2.2) determination, it was more specific.

Intestinal ALP, Biliary ALP, and Origin ALP

Intestinal ALP, when present in plasma, migrates in the beta-globulin position. It may be seen in health, particularly after a fatty meal, especially in blood group 0 and B secretor-positive individuals [5].

The demonstration of intestinal alkaline phosphatase in plasma has been recommended for differentiating intra-hepatic from extra-hepatic jaundice. The intestinal fraction was reported to be present in 45% of intra-hepatic jaundice but absent in extra-hepatic jaundice [6], and it was claimed that 'if ALP is increased in the jaundiced patient and an intestinal (I) band is present the lesion is likely to be intra-hepatic'. The reason suggested for this, is that bile salts solubilize ALP from the gut wall, their absence from the gut in extra-hepatic obstruction would prevent solubilization and absorption of the intestinal enzyme and result in reduction of its plasma activity. In contrast, parenchymal disease destroys receptors for intestinal phosphatase on the liver cell surface, so that this fraction would not be taken up by the liver and its plasma level would increase [7].

This claim was investigated in 47 consecutive jaundiced patients, 33 with intra-hepatic and 14 with extra-hepatic jaundice [8]. The presence of an I band in plasma samples separated on cellulose acetate had a sensitivity for intra-hepatic jaundice of 42%, with a specificity of 93%, figures similar to that quoted above. The positive predictive value was 93% and the negative predictive value 41%. However, the overall accuracy (efficiency) was only 57%, so clearly the test has considerable limitations.

Two other ALP isoenzymes known to be frequently elevated in cholestatic jaundice, alpha-1, biliary ALP and origin band ALP, were also examined in these patients. The alpha-1 ALP could not distinguish between intra-hepatic and extra-hepatic jaundice. It had a sensitivity of 64% for extra-hepatic disease, but poor specificity of only 48%, so that overall accuracy was low at 53%. The origin band had some differential value, but was rather insensitive, with a sensitivity of 57%. It showed a specificity of 76%, however, and an accuracy of 70%.

The differential diagnosis of jaundice was further examined by combined consideration of the I band, which has high specificity for intra-hepatic jaundice, and the origin band, which has moderate specificity for extra-hepatic jaundice. Using the technique of logistic model-fitting statistical analysis, it was observed that when the I band was present and the origin band absent, there was a 97% probability of intra-hepatic jaundice. With an absent I band but with origin ALP, the probability fell to 34%. Though ALP isoenzyme findings do have some differential diagnostic value in jaundice, clearly it would be foolhardy to rely on this over-much.

'Liver-Like' ALP in Transient Hyperphosphatasaemia of Infancy

An isoenzyme of mobility between biliary and liver phosphatase may appear in plasma in the condition of transient hyperphosphatasaemia of infancy [9, 10]. This increasingly recognised condition is characterized by the accidental finding of a markedly elevated plasma alkaline phosphatase, frequently up to 20 times the upper reference limit. This is transient, generally of less than 3 months' duration, occurs particularly in children under 3 years, is not due to clinical bone or liver disease and is unaccompanied by elevation of plasma gamma glutamyltransferase (GGT) or other liver enzymes.

Fifteen children (9 males and 6 females) with this disorder were studied. All were less than 3 years of age and showed transiently increased total

ALP activity as an incidental finding. A notable feature was the high incidence of accompanying gastrointestinal disorder or virus infection [11, 12]. Because of this, it was suggested that the hyperphosphatasaemia might be an unusual response to virus infection, particularly of the gastrointestinal tract.

The alkaline phosphatase isoenzyme electrophoretic pattern in this disorder was very distinctive and this facilitated diagnosis. On cellulose acetate, a staining band of mobility intermediate between alpha-1 and liver ALP was seen, together with a second slower band with the mobility of bone ALP [11]. The anodal band had heat-stability similar to liver ALP whereas the slower band showed the heat-lability characteristic of bone isoenzyme [12].

The effect of brief pre-incubation with neuraminidase on the electrophoretic mobility of the ALP isoenzymes in this condition was examined [12]. Incubation times insufficient to affect the mobility of biliary or normal liver ALP in control plasmas caused marked retardation of the anodal band in transient hyperphosphatasaemia, so that it moved to the position normally occupied by liver ALP. Since neuraminidase removes sialic acid groups from ALP, this marked mobility change of the anodal band suggests that this band represents excessively sialylated liver ALP ('liver-like' ALP), though the reason for this is unknown.

Immunoglobulin-Bound ALP

Enzymes in plasma may be bound by naturally occurring immunoglobulins to give enzyme-immunoglobulin complexes. Such binding may result in unexpected alteration of plasma total enzyme activity and unusual isoenzyme electrophoretic patterns. ALP-binding immunoglobulins in plasma are observed in up to 0.5% of patients, especially in patients over 60 years of age. They are found with increased frequency in patients with autoimmune disease [13] and in 15% of patients with ulcerative colitis [14]. Their unexpected occurrence in a patient's plasma should therefore alert the observer to the possible presence of these disorders. The binding immunoglobulin (Ig) is of IgG class in 90% and of lambda type in two-thirds. Almost all complexes bind liver and bone ALP, but not the immunologically distinct alpha-1, intestinal or placental phosphatase.

As a consequence of immunoglobulin binding, the electrophoretic mobility of bound liver and bone phosphatase is retarded, and staining is observed in the beta-globulin position adjacent to the position normally occupied by intestinal phosphatase. Ig-bound ALP may be confused with intestinal ALP because of its similar electrophoretic mobility. Distinguishing features found to assist their differentiation were the absence of staining in the position normally occupied by the liver and bone isoenzymes, binding of liver and bone phosphatase added in vitro, liberation of bound phosphatase by proteolysis, and immunofixation of the complex with immunoglobulin-specific antisera. Additionally, unlike intestinal ALP, the Ig-ALP complexes were generally resistant to *L*-phenylalanine inhibition, but retarded by neuraminidase.

Hepatoma-Specific Phosphatase

So-called 'hepatoma-specific' ALP is an onco-fetal protein which migrates rapidly on electrophoresis in advance of liver phosphatase. It has been reported to be present in the plasma of 18% of United Kingdom patients with hepato-cellular carcinoma [15]. Plasma from 10 patients with hepatitis B-negative, hepato-cellular carcinoma (7 with cirrhosis and alpha-fetoprotein increase in plasma) were studied by electrophoresis [Rosalki, Foo, Dash, Thomas, unpubl.] but no sample showed hepatoma-specific isoenzyme. Clearly, therefore, in a United Kingdom population, the frequency of the enzyme is too low to be of much diagnostic value for this condition.

Conclusion

This communication, though restricted to the consideration of information that can be derived from ALP isoenzyme electrophoresis in conditions personally studied by the author, nevertheless demonstrates the remarkable range of clinical situations in which plasma alkaline phosphatase isoenzymes may be altered. Simple electrophoretic separation procedures readily carried out in the clinical laboratory, greatly extend the diagnostic value of plasma alkaline phosphatase determination.

Acknowledgements

The observations described have been made in collaboration with the colleagues indicated in the text or in the reference list, but special acknowledgement must be made of the valued assistance of Dr. A.Y. Foo in all the studies.

References

1 Rosalki, S.B.; Foo, A.Y.: Two new methods for separating and quantifying bone and liver alkaline phosphatase isoenzymes in plasma. Clin. Chem. *30:* 1182–1186 (1984).

2 De Broe, M.; Roels, F.; Nouwen, E.J.; Claeys, S.L.; Wieme, R.J.: Liver plasma membrane: the source of high molecular weight alkaline phosphatase in human serum. Hepatology *5:* 118–128 (1985).

3 Viot, M.; Joulin, C.; Cambin, P.; Knebs, B.P.; Schneider, M.; Lelanne, C.M.: The value of serum alkaline phosphatase α_1 isoenzyme. Biomedicine *31:* 74–77 (1979).

4 Karmen, C.; Mayne, P.D.; Foo, A.Y.; Parbhoo, S.; Rosalki, S.B.: Measurement of biliary alkaline phosphatase by minichromatography and by electrophoresis and its application to the detection of liver metastases in patients with breast cancer. J. clin. Path. *37:* 212–217 (1984).

5 Langman, M.J.S.; Leuthold, E.; Robson, E.B.; Harris, J.; Luffman, J.E.; Harris, H.: Influence of diet on the 'intestinal' component of serum alkaline phosphatase in people of different ABO blood groups and secretor status. Nature, Lond. *212:* 41–43 (1966).

6 Warnes, T.W.; Hine, P.; Kay, G.; Smith, A.: Intestinal alkaline phosphatase in the diagnosis of liver disease. Gut *18:* 274–278 (1977).

7 Warnes, T.W.; Hine, P.; Kay, G.; Smith, A.: Intestinal alkaline phosphatase in bile: evidence for an enterohepatic circulation. Gut *22:* 493–498 (1981).

8 Goold, M.F.; Foo, Y.; Mayne, P.D.; Rosalki, S.B.: Alkaline phosphatase isoenzymes in the differential diagnosis of jaundice. Abstr. Ass. Clinical Biochemists, National Meet., 1984, p. 42.

9 Posen, S.; Lee, C.; Vines, R.; Kilham, H.; Latham, S.; Keefe, J.F.: Transient hyperphosphatasemia of infancy. An insufficiently recognised syndrome. Clin. Chem. *23:* 292–294 (1977).

10 Wieme, R.J.: More on transient hyperphosphatasemia of infancy – an insufficiently recognised syndrome (Letter). Clin. Chem. *24:* 520–522 (1978).

11 Rosalki, S.B.; Foo, A.Y.: More on transient hyperphosphatasemia of infancy (Letter). Clin. Chem. *29:* 723 (1983).

12 Rosalki, S.B.; Foo, Y.: Transient hyperphosphatasemia of infancy. Four new cases and a suggested etiology (Letter). Clin. Chem. *26:* 1109 (1979).

13 Crofton, P.M.; Kilpatrick, D.C.; Leitch, A.G.: Complexes in serum between alkaline phosphatase and immunoglobulin G. Immunological and clinical aspects. Clinica chim. Acta *111:* 257–265 (1981).

14 Leroux-Roels, G.G.; Wieme, R.J.; De Broe, M.E.: Occurrence of enzyme-immunoglobulin complexes in chronic inflammatory bowel disease. J. Lab. clin. Med. *97:* 316–321 (1981).

15 Kay, P.; Johnson, P.J.; Warnes, T.W.; Williams, R.; Longson, M.; Laing, I.; Smith, A.; Lee, N.W.: Hepatoma-specific alkaline phosphatase (H-ALP) in high and low incidence areas of primary liver cancer. Gut *23:* AGO7 (1982).

S.B. Rosalki, MD, Chemical Pathology and Human Metabolism, The Royal Free Hospital and School of Medicine, Pond Street, London NW3 2QG (UK)

Adv. clin. Enzymol., vol. 3, pp. 19–29 (Karger, Basel 1986)

Alkaline Phosphatases Isoforms in Prenatal Diagnosis of Cystic Fibrosis and Hypophosphatasia

I. Maire, A. Vialle, M. Mathieu

Laboratoire d'Enzymologie, Hôpital Debrousse, Lyon, France

Alkaline phosphatases (ALP; EC 3.1.3.1) are esterases exhibiting low substrate specificities. They apparently derive from three different loci: the placental group (P-ALP), the intestinal group (I-ALP) and the non-placental non-intestinal group (bone, liver, kidney) (L-ALP). Furthermore, ALP are glycoprotein enzymes and variations in their carbohydrate contents may also contribute to their heterogeneity [1].

Cystic fibrosis (McKusick 21970) (CF) is the most common lethal or semilethal genetic disease in the Caucasian population with a prevalence of about 1:2,000 live births in the United States [2]; in Europe, the frequency of the disease is less in Sweden (1:8,000) and Italy (1:15,000) than in 'middle' European countries (England 1:2,400 and Germany 1:3,300). All studies have confirmed the hypothesis of autosomal recessive inheritance and estimates of the frequency of the heterozygotes (based on the assumption of a single gene mutation) suppose that over 4% of the Caucasian population carry the gene.

Much effort has been devoted over the last 20 years to the search for the biochemical defect that underlies this disease. Despite all these extensive investigations, the primary cause of the disease remains unknown and any proposed system for in utero detection of the disease is likely to be founded on a secondary characteristic. The first hopes [3, 4] of the possibility of prenatal diagnosis have been disappointing since they either did not give reproducible results [5] or gave rise to false-negative and false-positive results [6]. More recent attempts are based upon the observation that the activities of two microvillar enzymes (gamma-glutamyl transferase and aminopeptidase M) are reduced in amniotic fluid (AF) of fetuses affected with CF [7] but some overlap exists between the activities in normal and affected fetuses [8, 9]. The disaccharidases and the I-ALP were also depressed [11, 12]. The I-ALP decrease first demonstrated by an inhibition technique was confirmed by Brock et al. [13] with a specific monoclonal antibody. Several hypotheses have been made to explain the decreased activities of microvillar enzymes in CF AF: a failure of development of microvilli of CF fetuses was first suggested [7], but the possibility that the viscous meconium traps the enzyme from the fetal gut seems more likely [10, 14].

Hypophosphatasia (McKusick 24150) is a rare genetic disorder characterized by defective bone mineralization [15, 16]. Depending upon the age of onset and the severity of the disease, three forms are differentiated: the severe infantile type, childhood hypophosphatasia, and the adult form. Autopsy studies of infants affected with infantile hypophosphatasia have revealed striking ALP deficiency in bone, liver and kidney but normal ALP activities in intestine and placenta [17–20]. A decreased ALP activity has been demonstrated in fibroblasts from different types of hypophosphatasia [18, 19, 21], which is in good agreement that an L-ALP isoform exists in these cells [22]; nevertheless, Vanneuville and Leroy [23] found that only the ALP induced by starvation of the cells seems identical with the enzyme in liver. Prenatal diagnosis of the infantile form has been attempted in a limited number of cases [15, 22, 24–29] and the combination of two methods has proven to be useful: the ultrasound and radiographic examination of the fetus and the study of cultured AF cell ALP activity. The AF presents a normal total ALP activity [23, 24, 27] as expected since more than 80% of ALP activity in AF is of intestinal origin [22, 24], a few percent from the L-ALP group, with a smaller contribution by placental enzyme. The cultured AF cell ALP was found to be decreased in some affected fetuses [24, 26], but some authors noticed falsely high activity [28, 29]; these discrepancies could be explained by the fact that some samples of cultured AF cells may contain other ALP isoforms than the L-ALP group [22]. It is notable that in hypophosphatasia some residual activity different from the usual L-ALP (I-ALP or an embryonic ALP later replaced by the adult form of L-ALP) was found post-mortem in tissues such as liver, bone and kidney [20, 28, 30].

I-ALP and the Prenatal Diagnosis of CF

Among the difficulties of the prenatal diagnosis of CF, some may be related to the poor knowledge of the ontogenic development of ALP isoforms found in AF. The first systematic studies comparing fetal and adult forms of ALP [31, 32] found very similar properties (various chemical inhibitors, heat stability, immunoreactivity with polyclonal antibodies) between adult and fetal forms, except for the greater anodal electrophoretic mobility of the I-ALP fetal form and its retardation after neuraminidase treatment. Behrens et al. [33] concluded that fetal I-ALP was formed by 2 non-identical subunits: one resembling P-ALP; the other corresponding to a glycosylated subunit of the adult intestinal enzyme. Other recent studies using monoclonal antibodies [34, 35] or peptide mapping of human adult I-ALP [36, 37] suggested that adult and fetal I-ALP represent structurally distinct proteins; whether they are encoded by separated gene loci [37] or they are the result of differential processing of the product of the same gene may be discussed.

The prenatal diagnosis of CF is based upon the study of I-ALP in AF [11–13] and the inhibition studies with phenylalanine and homoarginine assumed an identical behaviour of adult and fetal I-ALP. To ascertain this fact, we prepared butanol extracts [38] from intestine of fetuses at different stages of gestation, and of adults who underwent surgical intestinal resection for neoplasm (using an unaffected part selected by the pathologist). ALP activity was measured using either methylumbelliferyl phosphate (4MU-P) as substrate by a modification of the method of Guibault and Reitz [39] or *p*-nitrophenylphosphate (PNPP) using the method recommended for serum [40]. By inhibition studies, only the I-ALP of very early gestational stages (before 17 weeks of amenorrhoea) was found to be different (table I). By high-performance liquid chromatography (HPLC) on an Ultropac TSK DEA-5PW column using an LKB HPLC pump and a NaCl gradient from 0 to 200 mmol/l, all the intestinal extracts were eluted as a single peak at the following Na^+ concentration: 150–160 mEq/l for 14- to 19-week fetuses, 140 mEq/l for those of 20 weeks, 134 mEq/l for those of 23 weeks and 104 mEq/l for adult intestine; after neuraminidase treatment, the 15-week fetal intestine ALP was resolved in several peaks, the main one being eluted at 36 mEq/l and the secondary ones at 75 and 180 mEq/l while the adult intestinal extract was unaffected by this treatment. By electrophoretic studies, 20-week fetal intestine ALP presented a greater anodal mobility reduced by neuraminidase treatment, as reported by others [31].

The inhibition studies were then applied to AF sampled between 16 weeks 5 days and 18 weeks of amenorrhoea using either PNPP or 4MU-P as substrate. Our retrospective study confirmed the existence of false-positives (3 among the 200 AF tested). In our prospective study, 4 cases of affected fetuses (among 14 'at risk' families) were diagnosed following the criteria shown in table II: *the results were more clear-cut between 17 weeks 4 days and 18 weeks 1 day* but were less informative at 16 weeks 3 days and after 19 weeks. Considering that I-ALP is the major constituent of AF ALP at this gestational stage, the lack of reliability of too early sampling may be explained by the results of inhibition studies reported in table I: they show that inhibition studies may lead to ambiguous results until 16 weeks of amenorrhoea since early intestinal ALP behaves differently. The decreased reliability of late samples may possibly be explained by the increased percentage of fetal urine in the composition of AF.

All the cases reported here led to a therapeutic abortion. In 2 cases, we were able to study the I-ALP of the fetus after therapeutic abortion; the I-ALP was found to be qualitatively normal either in intestine or in meconium (table II). Quantitatively, the I-ALP activity was found to be normal

Table I. ALP in intestine and meconium of fetuses at different stages of gestation

	4MU-P, % of inactivation of ALP activity by			PNPP, % of inactivation of ALP activity by		
	homoarginine 10 mmol/l	phenylalanine 5 mmol/l	bromotetramisole 0.1 mmol/l	homoarginine 10 mmol/l	phenylalanine 5 mmol/l	bromotetramisole 0.1 mmol/l
Intestine 14, 15 weeks	53, 51	65, 51	59, 63	43, 49	38, 36	45, 58
Intestine 16 weeks	–	–	–	34	34	–
Fetus G.A. 17 weeks, 4 days						
Intestine	21	76	28			
Meconium	18	77	22	18	55	14
Fetus G.O. 18 weeks, 1 day						
Intestine	18	78	21			
Meconium	10	74	15	13	55	8
Intestine 20, 23 weeks, adult	17, 27, 21	78, 78, 26	19, 29, 26	15, 8, 12	53, 50, 55	24, 12, 6

in 1 case (Mrs. G.O.) and decreased in the other case (Mrs. G.A.) but the meconium was very adherent and could have retained intestinal cells leading to a loss of activity; furthermore, the higher ALP activity in this meconium makes this hypothesis very likely. Nevertheless, the presence of this meconium is not a constant finding [13] and the decreased activity of I-ALP in AF remains to be explained in these cases.

L-ALP and Prenatal Diagnosis of Hypophosphatasia

The reliable prenatal diagnosis of hypophosphatasia supposes a knowledge of the ontofetal development of the L-ALP and of the isoenzyme content of cultured AF cells. No differences were found by Miki et al. [31] between adult and fetal (18–36 weeks) L-ALP. In contrast, Lehmann [41] differentiates 3 different bone ALP: the 'early' fetal (18–23 weeks of gesta-

Table II. CF prenatal diagnosis

	Gestational age	ALP (substrate: 4MU-P)			ALP (substrate: PNPP)		
		activity nkat/l	% inhibition		activity nkat/l	% inhibition	
			homoarginine 10 mmol/l	phenylalanine 5 mmol/l		homoarginine 10 mmol/l	phenylalanine 5 mmol/l
Controls							
16 weeks 5 days → 18 weeks 2 days (n = 200)							
Median		68	21	67.5	340	16.5	59
5th or 95th percentile		>21	<40	>58	>123	<35.5	>42
20 weeks → 22 weeks (n = 80)							
Median		31	28	59			
Mrs. G.A.	16 weeks 3 days	86	42	56	473	33	39
	17 weeks 4 days	57	56	46	316	49	26
Mrs. G.O.	17 weeks 1 day	23	57	43	175	50	19
	18 weeks 1 day	43	74	38	191	58	12
Mrs. V.A.	17 weeks	14	46	53			
	18 weeks	16	63	48			
Mrs. R.U.	18 weeks	16.8	63	28	80	61	26
	19 weeks 2 days	9.2	58	45	37	27	22

tion), the 'late' fetal (7–9 months of gestation) and the adult form; they may be distinguished by slightly different electrophoretic mobility; furthermore, the 'early' fetal bone ALP shares common antigenic determinants with adult bone ALP (and thus with L-ALP) but presents antigenic determinants not found in liver or kidney tissues: these additional antigenic determinants are shared with term P-ALP and adult I-ALP.

We studied ALP in liver and bone of fetuses at different stages of gestation in comparison with those of a newborn who died from hypophosphatasia and with adult tissues (table III). In livers, no differences could be found, either by inhibition studies or by HPLC elution, whatever the gestational age. In the liver of the hypophosphatasic infant, the very low residual activity presented an abnormal response to the inhibitors, and a surprising activating effect of phenylalanine was noticed. In a 20-week hypophosphatasic fetus, we found strongly reduced ALP activity in liver (4%), kidney (0%) and brain (10%) by comparison with a 20-week non-hypophosphatasic fetus, and electrophoresis on cellulose acetate revealed a slightly retarded ALP responsible for residual activity in liver and brain. In bones, the inhibition studies did not show any difference, even in hypophospha-

Table III. Bone and liver ALP

	Substrate: 4MU-P % of inhibition of ALP by		Substrate: PNPP, % of inhibition of ALP by		HPLC mEq/l of Na^+ for elution of the peak
	homoarginine 10 mmol/l	phenylalanine 5 mmol/l	homoarginine 10 mmol/l	phenylalanine 5 mmol/l	
Liver					
15 weeks	88	26	75	8	160
20 weeks	87	16	74	25	161
Adult	88	27	74	15	156
Hypophosphatasia (Gla.)	64	+14[1]	–	–	155
Bone					
12 weeks	88	34	74	8	192
14 weeks	87	29	75	9	173
20 weeks	85	26	74	2	175
Adult	86	33	75	4	164
Hypophosphatasia (Gla.)	83	39	70	14	174

[1] % of activation.

tasic bone where we found only a striking decrease of activity. In the 20-week hypophosphatasic fetus, the ALP activity in bone was not detectable. HPLC elution of bone ALP extracts exhibited variations in concentrations of Na^{+} elution peak following the gestational age.

In AF of 2 hypophosphatasic fetuses, the ALP activity was normal in 1 case and decreased in the other; the reason of the decreased activity in 1 case can be hardly understood, considering that 17–18 weeks AF contain mainly I-ALP which was found to be normal or increased in our hypophosphatasic fetuses.

The ALP activity varied very widely in cultured AF cells of these fetuses (table IV). The cellular heterogeneity of the AF or the gestational age

Table IV. ALP in fibroblasts and AF cells

Cells studied	ALP (substrate 4MU-P) μkat/kg of cell proteins	% of inhibition: – % of activation: + of ALP activity by homoarginine 10 mmol/l	phenylalanine 5 mmol/l	theophylline 10 mmol/l
Cultured AF cells				
Controls				
17 → 18 weeks (20)				
primary culture → sub 3	11 → 163	–15 → –85	–2 → –57	–23 → –96
16 weeks (1), sub 2	11.8	0	+11.5	+15
Hypophosphatasia				
An. (primary culture, sub 1)	1.9, 2.7	–	–	–
El. (sub 1, sub 1)	6.9, 9.8	(–, +13)	(+13, +16)	(+16, +37)
Cultured fibroblasts				
Controls				
Sub 4 → sub 8 (20)	18 → 125	–36 → –75	–4 → –39	–22 → –87
Hypophosphatasia				
El. (sub 8, sub 10, sub 3)	1.6, 2.0, 10.8	(–, –, –9)	(–, –, +19)	(–, –, +19)
Heterozygotes				
Mr. and Mrs. An.				
(primary culture, sub 5)	7.4 → 1.6			
Mr. and Mrs. Gla.				
(sub → sub 5)	28 → 14	(–68 → –28)	(–29 → –2)	(–73 → –11)

do not explain these wide variations of activity which are of the same order of magnitude in fibroblasts and are more probably related to culture conditions. In such conditions, the diagnosis of the heterozygotes in fibroblasts seems quite impossible (table IV). More interesting is the abnormal behaviour of the hypophosphatasic fibroblast enzyme, on which theophylline and phenylalanine exhibit an activating effect also shown by cultured AF cell ALP of an affected fetus in the same family. This surprising effect was also found once in an early sample (16 weeks) of AF cells and in the liver of a hypophosphatasic infant (table III), at least for phenylalanine. All these results suggest the presence of different forms of ALP, possibly fetal forms responsible in some cases for the residual activity in hypophosphatasic fetal liver [30] and of the electrophoretic variant described in some cultured AF cells [22].

A new biological material is now available for prenatal diagnosis: the chorionic villi obtained at the 10th week of amenorrhoea. Though considered not informative for prenatal diagnosis of hypophosphatasia, a low abnormal ALP was found in one case in the amnion of an hypophosphatasic fetus [42]. Using uncultured and cultured trophoblast, the percentages of inhibition by homoarginine and phenylalanine were found to be, respectively, 25 and 67% in 12-week uncultured material and 78 and 28% in 2 strains of 10 weeks cultured trophoblast cells, but the isoenzyme content of this material needs more investigation. Nevertheless, the presence of an embryonic enzyme in the trophoblast is more likely than the presence of a mixture of I-ALP and L-ALP in this tissue [43].

Conclusion

At the present time, all the prenatal diagnoses based on the study of ALP isoforms are limited by the poor knowledge of the ontofetal development of ALP at early stages of life and much basic effort in this field is necessary to increase the reliability of biochemical prenatal diagnosis of hypophosphatasia. For the prenatal diagnosis of CF, determination of I-ALP in AF by inhibition seems the best possibility if AF sampling is performed between 17.5 and 18.5 weeks; but the possibility of false-positives limits its application to 'at-risk' families and the rare possibility of false-negatives implies a very careful genetic counselling in 'at-risk' families.

References

1 Moss, D.: Alkaline phosphatases isoenzymes. Clin. Chem. *28:* 2007–2016 (1982).

2 Nadler, M.; Rao, G.; Taussig, L.: Cystic fibrosis; in Standbury, The metabolic basis of inherited disease; 4th ed., pp. 1683–1706 (McGraw, New York 1978).

3 Hösli, P.; Erickson, R.; Vogt, E.: Prospects for prenatal diagnosis of cystic fibrosis. Induction of biochemical abnormalities in fibroblasts from patients with cystic fibrosis by a urinary glycoprotein. Biochem. biophys. Res. Commun. *73:* 209–216 (1976).

4 Nadler, H.; Walsh, M.: Prenatal detection of cystic fibrosis on amniotic fluid. Lancet *ii:* 96–97 (1980).

5 Aitken, D.; Hoogeveen, A.: Alkaline phosphatase activity of normal and cystic fibrosis fibroblasts. J. med. Genet. *17:* 187–190 (1980).

6 Travers, H.; Schwartzman, M.; Benke, P.: False positive prenatal diagnosis of cystic fibrosis by protease assay. Am. J. Obstet. Gynec. *146:* 338–339 (1983).

7 Carbarns, N.; Gosden, C.; Brock, D.: Microvillar peptidase activity in amniotic fluid: possible use in the prenatal diagnosis of cystic fibrosis. Lancet *i:* 329–331 (1983).

8 Brock, D.; Bedgood, D.; Hayward, C.: Prenatal diagnosis of cystic fibrosis by assay of amniotic fluid microvillar enzymes. Hum. Genet. *65:* 248–251 (1984).

9 Aitken, D.; Yaqoob, M.; Ferguson-Smith, M.: Microvillar enzyme analysis in amniotic fluid and the prenatal diagnosis of cystic fibrosis. Prenat. Diagn. *5:* 119–127 (1985).

10 Van Diggelen, O.; Janse, H.; Kleijer, W.: Disaccharidases in amniotic fluid as possible markers for cystic fibrosis. Lancet *i:* 817 (1983).

11 Brock, D.: Amniotic fluid alkaline phosphatase isoenzymes in early prenatal diagnosis of cystic fibrosis. Lancet *ii:* 941–943 (1983).

12 Muller, F.; Berg, S.; Frot, J.; Boué, J.; Boué, A.: Prenatal diagnosis of cystic fibrosis. Prospective study of 51 pregnancies. Prenat. Diagn. *5:* 97–108 (1985).

13 Brock, D.; Barron, L.; Bedgood, D.; Van Heyningen, V.: Prenatal diagnosis of cystic fibrosis using a monoclonal antibody specific for intestinal alkaline phosphatase. Prenat. Diagn. *4:* 421–426 (1984).

14 Brock, D.: A comparative study of microvillar enzyme activities in the prenatal diagnosis of cystic fibrosis. Prenat. Diagn. *5:* 129–134 (1985).

15 Rasmussen, H.: Hypophosphatasia; in Stanbury, The metabolic basis of inherited disease; 5th ed., pp. 1497–1507 (McGraw, New York 1983).

16 Fallon, M.; Teitelbaum, S.; Weinstein, R.; Goldfischer, S.; Brown, D.; Whyte, M.: Hypophosphatasia: clinicopathologic comparison of the infantile, childhood, and adult forms. Medicine *63:* 12–24 (1984).

17 Danovitsch, G.; Baer, P.; Laster, L.: Intestinal alkaline phosphatase activity in familial hypophosphatasia. New Engl. J. Med. *278:* 1253–1260 (1968).

18 Brydon, W.; Crofton, P.; Smith, A.; Barr, D.; Harkness, R.: Hypophosphatasia: enzyme studies in cultured cells and tissues. Biochem. Soc. Trans. *3:* 927–929 (1975).

19 Vanneuville, F.; Leroy, L.: Enzymatic diagnosis of congenital lethal hypophosphatasia in tissues, plasma, and diploid skin fibroblasts. J. Inher. Metab. Dis. *4:* 129 (1981).

20 Mueller, H.; Stinson, R.; Mohyuddin, F.; Milne, J.: Isoenzymes of alkaline phosphatase in infantile hypophosphatasia. J. Lab. clin. Med. *102:* 24–30 (1983).

21 Whyte, M.; Vrabel, L.; Schwartz, T.: Alkaline phosphatase deficiency in cultured skin fibroblasts from patients with hypophosphatasia. Comparison of the infantile, childhood, and adult forms. J. clin. Endocr. Metab. *57:* 831–837 (1983).
22 Shin-Buerhing, Y.; Santer, R.; Osang, M.; Schaub, J.: Properties of human alkaline phosphatase and prenatal diagnosis of hypophosphatasia. J. Inher. Metab. Dis. *4:* 125–126 (1981).
23 Vanneuville, F.; Leroy, J.: Properties of alkaline phosphatase in crude homogenates of human diploid fibroblasts. Archs int. Physiol. Biochim. *88:* 155–161 (1980).
24 Mulivor, R.; Mennuti, M.; Zackari, E.; Harris, H.: Prenatal diagnosis of hypophosphatasia genetic, biochemical, and clinical studies. Am. J. hum. Genet. *30:* 271–282 (1978).
25 Hösli, P.; Rudd, N.: Alkaline phosphatase induction in fibroblast cultures. Prenatal diagnosis and carrier detection in hypophosphatasia. Monogr. hum. Genet. *10:* 70–74 (1978).
26 Kousseff, B.; Mulivor, R.: Prenatal diagnosis of hypophosphatasia. Obstet. Gynec. *57:* 9s–12s (1981).
27 Rudd, N.; Miskin, M.; Hoar, D.; Benzie, R.; Doran, I.: Prenatal diagnosis of hypophosphatasia. New Engl. J. Med. *295:* 146–148 (1976).
28 Blau, K.; Rattenbury, J.; Pryse-Davies, J.; Clark, P.; Standler, M.: Prenatal detection of hypophosphatasia. Cytological and genetic considerations. J. Inher. Metab. Dis. *1:* 37–39 (1978).
29 Vanneuville, F.; Leroy, J.; Van Elsen, A.: Alkaline phosphatase in cultured amniotic fluid cells: implications for prenatal diagnosis of hypophosphatasia. J. Inher. Metab. Dis. *5:* 39–40 (1982).
30 Gutman, A.; Milwidsky, A.: Hypophosphatasia in utero with normal activity in the liver. Clin. Biochem. *16:* 368 (1983).
31 Miki, K.; Oda, T.; Suzuki, H.; Niwa, H.; Endo, Y.; Lino, S.; Miyasaki, J.; Hirano, K.; ugiura, M.: Human fetal organ alkaline phosphatases. Clinica chim. Acta *85:* 115–124 (1978).
32 Mulivor, R.; Hanning, V.; Harris, H.: Developmental change in human intestinal alkaline phosphatase. Proc. natn. Acad. Sci. USA *75:* 3909–3912 (1978).
33 Behrens, C.; Enns, C.; Sussman, H.: Characterization of human foetal intestinal alkaline phosphatase. Biochem. J. *211:* 553–558 (1983).
34 Gogolin, K.; Wray, L.; Slaughter, C.; Harris, H.: A monoclonal antibody that reacts with non-allelic enzyme glycoproteins. Science *126:* 59–61 (1982).
35 Vockley, J.; Meyer, L.; Harris, H.: Differentiation of human adult and fetal intestinal alkaline phosphatases with monoclonal antibodies. Am. J. hum. Genet. *36:* 987–1000 (1984).
36 Vockley, J.; D'Souza, M.; Foster, C.; Harris, H.: Structural analysis of human adult and fetal alkaline phosphatases by cyanogen bromide peptide mapping. Proc. natn. Acad. Sci. USA *81:* 6120–6123 (1984).
37 Mueller, H.; Leung, H.; Stinson, R.: Different genes code for alkaline phosphatases from human fetal and adult intestine. Biochem. biophys. Res. Commun. *126:* 427–433 (1985).
38 Smith, I.; Lightstone, P.; Perry, J.: Separation of human tissue alkaline phosphatases by electrophoresis on acrylamide disc gels. Clinica chim. Acta *19:* 499–505 (1968).

39 Guibault, C.; Rietz, B.: Fluorometric assay of serum acid for alkaline phosphatase either in solution or on a semisolid surface. Clin. Chem. *21:* 1791–1794 (1975).
40 Commission enzymologie de la SFBC: Recommandations pour la mesure de la concentration catalytique des phosphatases alcalines dans le sérum humain à 30 °C. Annls Biol. clin. *40:* 86–164 (1982).
41 Lehmann, F.: Human fetal bone alkaline phosphatases. Preliminary immunologic studies. Oncodev. Biol. Med. *2:* 11–20 (1981).
42 Sakiyama, T.; Owada, N.; Kitagawa, T.: Hypophosphatasia: characterization of alkaline phosphatase in liver and its retrospective speculation for amniotic enzyme. Poster Presentation, Int. Symp. Inborn Errors of Metabolism in Human, Interlaken 1980.
43 Maxwell, D.; Blau, K.; Johnson, R.; Lilford, R.: Activities of alkaline phosphatase in first trimester chorion biopsy tissue. Prenat. Diagn. *5:* 283–286 (1985).

I. Maire, MD, Hôpital Debrousse,
29 Rue Soeur Bouvier, F–69322 Lyon Cedex 5 (France)

Adv. clin. Enzymol., vol. 3, pp. 30–38 (Karger, Basel 1986)

Cancer-Associated Forms of Human Placental-Type Alkaline Phosphatase

P.J. McLaughlin

Pregnancy Immunology Group, Department of Immunology,
University of Liverpool, UK

Human alkaline phosphatase (AP) isoenzymes are dimeric glyco- or sialoglycoproteins usually recognized by their ability to hydrolyse phosphoester bonds at alkaline pH. However, their physiological role may be in hydrolyzing phosphoproteins at neutral pH [18, 50]. This action could regulate a variety of cellular functions (including transport of molecules across membranes and regulation of cell growth, differentiation and division) through interaction with enzymes, growth-factor receptors and inositol phospholipids [23]. There appear to be three main isoenzymic forms of human alkaline phosphatase, coded for by at least three different gene loci: intestinal alkaline phosphatase (IAP), placental alkaline phosphatase (PLAP) and liver/bone/kidney type or 'tissue non-specific' AP.

Normal and Neoplastic Forms of Tissue Non-Specific AP

The tissue non-specific forms of AP are thought to be the products of one genetic locus but may differ from each other in co- or post-translational glycosylation and sialysation [37]. They form the major AP activity in a number of organs and tissues, and comprise the major or sole AP activity in the serum in healthy non-pregnant adults. Tumour cells may express quantities of AP but because of the ubiquity of this enzyme it has not proved to be a useful tumour marker.

Normal and Neoplastic Forms of IAP

The AP of human small intestine occurs as two forms: fetal and adult IAP. These may represent products of separate gene loci [38, 57]. A hepatoma-associated product, the Kasahara enzyme, apparently related to the fetal IAP, has been proposed as an oncofetal marker [16]. This has a limited tumour distribution however, with circulating enzyme being found regularly only in 15% of patients with hepatoma [14].

Non-Neoplastic Forms of Placental-Type AP (PLAP)

PLAP is a major sialoglycoprotein of term placental syncytiotrophoblast plasma membranes and is also released into the maternal peripheral circulation after the first trimester of pregnancy. The enzyme is encoded for at a single locus, but there is a high degree of allelic polymorphism [6, 44]. PLAP can be distinguished from the other isozymic forms (IAP and AP) by its remarkable heat stability, by differential sensitivity to inhibition with certain *L*-amino acids and peptides, and by reactivity with specific antisera or monoclonal antibodies (mAbs) [10, 30, 33, 47]. Similar or identical enzymic forms are expressed in small amounts in some adult tissues [13, 31, 58]. A related but distinct isoenzyme has been described: the placental-like alkaline phosphatase (PL-like AP). This is most readily distinguished from PLAP by its sensitivity to inhibition by *L*-leucine [4] and non-reactivity with certain anti-PLAP mAbs [31, 32]. The PL-like AP has been suggested to be a product of a separate genetic locus from that coding for PLAP [34] or it may merely represent co- or post-translational structural modifications of the PLAP polypeptide chain. This *L*-leucine-sensitive PL-like AP form is found in normal testicular and thymic tissue [13, 34, 58]. Two further normal tissues, lung and cervix, express concurrently both PLAP and PL-like AP isoenzymes [41, 58]. However, these enzymes are not normally released into the peripheral circulation in detectable quantities except in cigarette smokers, who usually have circulating PL-like AP (but not PLAP) probably derived from damaged lung alveolar tissue [32, 58].

Neoplastic Forms of PLAP

The neoplastic forms of PLAP and PL-like AP originally were termed Regan and Nagao isozymes, respectively [11, 39]. It is probable that these are products of the same locus/loci as PLAP and PLAP-like AP, differing perhaps in post-translational processing [20]. These PLAPs are proving to be more generally applicable oncofetal markers than the Kasahara enzyme, being found both in a number of different tumour types and in a high proportion of some cancers, particularly those of the human reproductive tract [48]. Detection of the placental-type separately from other forms of human AP may be effected by heat-inactivation and/or amino acid inhibition, by use of polyclonal xenoantisera or, more recently, by application of mouse hybridomas producing specific mAbs reactive with human PLAP forms.

These various methods have been applied to detection of inappropriately expressed PLAP and PL-like AP in cancer patients. The approaches have been: (1) Detection of the marker in patients' sera for use as a screening procedure and in follow-up. (2) Detection of the marker in tumour tissue either using histochemical and immunohistological techniques or by assaying placental-type AP in solubilized tissue extracts. (3) Detection of tumour deposits in vivo using radiolabelled monoclonal antibodies.

Serological Assays for PLAP/PL-like AP

Although circulating PLAP occurs in a proportion of a variety of tumour types [12, 30, 56], its main clinical application has been in ovarian and testicular neoplasms. Detectable circulating enzyme has been found in 30–60% of ovarian cancer patients [5, 15, 24, 28, 43]. However, the usefulness of these findings has been limited by the detection of enzyme in the sera of healthy individuals [5], particularly cigarette smokers [26, 53]. Recently, specific monoclonal antibodies have been developed which not only distinguish the placental from other forms of AP but also recognise variants within the PLAP group [33, 46]. One such mAb (H317) has proved particularly useful in that it recognises PLAP but not the similar PL-like AP form [29–32]. When used in tandem with a mAb (H17-E2) which recognizes all forms of PLAP and PL-like AP, it is possible to separately quantitate these two enzymic forms. This is important for serological screening because the form circulating in some healthy individuals and most cigarette smokers is the H317-unreactive PL-like AP form [32] and so does not give false-positive readings in the H317-based assay for PLAP in cancer patients. Hence, this assay is more specific for identifying cancer-associated serum PLAP than assays unable to distinguish the PLAP from PL-like AP forms. However, some sensitivity is lost for those tumours which may release PL-like AP into the circulation. This is not a particular problem in

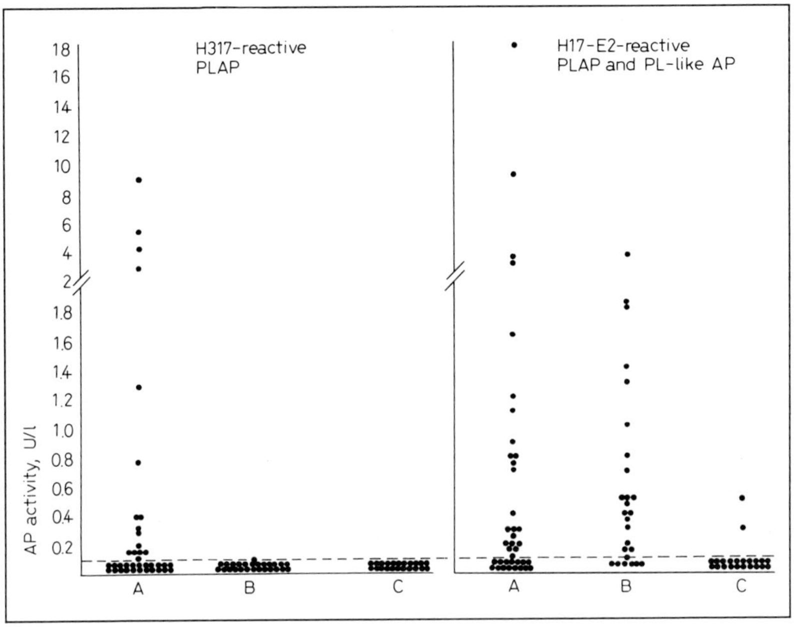

Fig. 1. Serum PLAP and PL-like AP. A = Ovarian cancer patients (n = 38); B = healthy cigarette smokers (n = 26); C = healthy non-smokers (n = 22).

ovarian cancer as most tumours release PLAP either alone or together with PL-like AP (fig. 1) but in seminoma, PL-like AP rather than PLAP is released into the blood [21, 29, 35] and thus the specificity of the assay for cancer detection is limited by the common occurrence of serum PL-like AP in healthy cigarette smokers [17, 36, 54]. However, as the prevalence of circulating PL-like AP in active seminoma may approach 100% [8] the test has a useful negative predictive value.

Detection of PLAP in Tumour Tissue

PLAP expression by tumours can be demonstrated histochemically [45] and immunohistologically using both polyclonal [19, 55] and monoclonal antibodies [25, 27, 28, 40, 42, 49]. The highest incidence of detection has been in testicular tumours, although many ovarian neoplasms also express this marker. Positive immunohistological findings have been reflected in the quantitation of PLAP in solubilized tissue extracts [1]. In ovarian tumours, PLAP levels may increase with the malignancy of the tumour although detected levels are variable [5, 28, 43]. In testicular tumours, particularly seminoma, levels of PL-like AP may be up to two orders of magnitude greater than the upper limit for the normal testis [21].

Apart from quantitative differences in PLAP and PL-like AP expression by normal and tumour tissues, there is also evidence of qualitative differences in expression. For example, it has been suggested that a significant proportion of PLAP in tumour cells may be intracellular [45, 52] and that extra antigenic sites may be exposed on the PLAP expressed by tumours [51].

Detection of Cancer in vivo

Radiolabelled murine IgG_1 monoclonal antibody H317, reactive with PLAP, has been shown to localize preferentially in resected ovarian tumour rather than normal tissue, giving an average tumour:non-tumour localization ratio of 40:1 [9]. This, and other, PLAP-reactive, ^{123}I-labelled mAbs have now been used successfully to localize tumour in vivo in patients with ovarian and testicular cancer. Clinically unsuspected tumour recurrence was revealed in some cases [2, 3, 22]. The radioisotope ^{111}In has also been used for labelling PLAP-reactive mAbs which have localized primary and metastatic tumours of the ovary, testis and cervix [7]. These results suggest the possibility of tumour treatment with mAb-targetted radiotherapy or cytotoxic drug therapy.

Acknowledgements

I am grateful to the Imperial Cancer Research Fund and the North West Cancer Research Fund for support. The skilful secretarial help of Miss Elizabeth Thompson is much appreciated.

References

1 Benham, F.J.; Andrews, P.W.; Knowles, B.B.; Bronson, D.L.; Harris, H.: Alkaline phosphatase isoenzymes as possible markers of differentiation in human testicular teratocarcinoma cell lines. Devl Biol. *88:* 279–287 (1981).

2 Critchley, M.; Brownless, S.; Patten, M.; McLaughlin, P.J.; Tromans, P.M.; McDicken, I.W.; Johnson, P.M.: Radionuclide imaging of epithelial ovarian tumours with ^{123}I-labelled monoclonal antibody (H317). Clin. Radiol. *37:* 107–112 (1986).

3 Davies, J.O.; Davies, E.R.; Howe, K.; Jackson, P.C.; Pitcher, E.M.; Sadowski, C.S.; Stirrat, G.M.; Sunderland, C.A.: Radionuclide imaging of ovarian tumours with

^{123}I-labelled monoclonal antibody ($NDOG_2$) directed against placental alkaline phosphatase. Br. J. Obstet. Gynaec. *92:* 277–286 (1985).
4 Doellgast, G.J.; Fishman, W.H.: Inhibition of human placental-type alkaline phosphatase variants by peptides containing *L*-leucine. Clinica chim. Acta *75:* 449–454 (1977).
5 Doellgast, G.J.; Homesley, H.D.: Placental-type alkaline phosphatase in ovarian cancer fluids and tissues. Obstet. Gynec. *63:* 324–329 (1984).
6 Donald, L.J.; Robson, E.B.: Rare variants of placental alkaline phosphatase. Ann. hum. Genet. Lond. *37:* 303–317 (1974).
7 Epenetos, A.A.; Hooker, G.; Durbin, H.; Bodmer, W.F.; Snook, D.; Begent, R.; Oliver, R.T.D.; Lavender, J.P.: Indium-III labelled monoclonal antibody to placental alkaline phosphatase in the detection of neoplasms of testis, ovary and cervix. Lancet *ii:* 350–353 (1985).
8 Epenetos, A.A.; Munro, A.J.; Tucker, D.F.; Gregory, W.; Duncan, W.; MacDougall, R.H.; Faux, M.; Travers, P.; Bodmer, W.F.: Monoclonal antibody assay of serum placental alkaline phosphatase in the monitoring of testicular tumours. Br. J. Cancer *51:* 641–644 (1985).
9 Epenetos, A.A.; Snook, D.; Halnan, K.; Lambert, J.; Field, S.B.; MacGregor, W.; Wood, C.; Orr, J.S.; Durbin, H.; Taylor-Papadimitriou, J.; Bodmer, W.F.; McLaughlin, P.J.; Johnson, P.M.: Limitations of radiolabelled monoclonal antibodies for localization of human neoplasms. Cancer Res. (in press, 1986).
10 Fishman, W.H.: Perspectives on alkaline phosphatase isoenzymes. Am. J. Med. *56:* 617–650 (1974).
11 Fishman, W.H.; Inglis, N.R.; Stolbach, L.L.; Krant, M.J.: A serum alkaline phosphatase isoenzyme of human neoplastic cell origin. Cancer Res. *28:* 150–154 (1968).
12 Fishman, W.H.; Inglis, N.R.; Vaitukaitis, J.; Stolbach, L.L.: Regan isoenzymes and HCG in ovarian cancer. Natn. Cancer Inst. Monogr. *42:* 63–73 (1975).
13 Goldstein, D.J.; Rogers, C.; Harris, H.: A search for trace expression of placental-like alkaline phosphatase in non-malignant human tissues. Clinica chim. Acta *125:* 63–75 (1982).
14 Hada, T.; Amuro, Y.; Higashino, K.: Studies of intestinal-like alkaline phosphatase (the Kasahara variant); in Stigbrand, Fishman, Human alkaline phosphatases, pp. 235–242 (Liss, New York 1984).
15 Haije, W.G.; Meerwaldt, J.H.; Talerman, A.; Kuipers, T.J.; Baggerman, L.; Teeuw, A.H.; van Der Pompe, W.B.; van Driel, J.: The value of a sensitive assay of carcinoplacental alkaline phosphatase (CPAP) in the follow-up of gynaecological cancers. Int. J. Cancer *24:* 288–293 (1979).
16 Higashino, O.K.; Otani, R.; Kuda, S.; Yamamura, Y.: A fetal intestinal-type alkaline phosphatase in hepatocellular carcinoma tissue. Clin. Chem. *23:* 1615–1623 (1977).
17 Horwich, A.; Tucker, D.F.; Peckham, M.J.: Placental alkaline phosphatase as a tumour marker in seminoma using the H17-E2 monoclonal antibody assay. Br. J. Cancer *51:* 625–629 (1985).
18 Huang, K.-P.; Robinson, J.C.; Chou, J.Y.: Phosphoprotein phosphatase activity associated with human placental alkaline phosphatase. Biochem. biophys. Res. Commun. *70:* 186–192 (1976).
19 Jacobsen, G.K.; Norgaard-Pedersen, B.: Placental alkaline phosphatase in testicular

germ cell tumours and in carcinoma-in-situ of the testis. Acta Pathol. microbiol. immunol. scand. A, Pathol. *92:* 323–329 (1984).
20 Jemmerson, R.; Shah, N.; Fishman, W.H.: Evidence for homology of normal and neoplastic placental alkaline phosphatases as determined by monoclonal antibodies to cancer-associated enzyme. Cancer Res. *45:* 3268–3273 (1985).
21 Jeppson, A.; Wahren, B.; Brehmer-Anderson, E.; Silfversward, C.; Stigbrand, T.; Millan, J.L.: Eutopic expression of placental-like alkaline phosphatase in testicular tumours. Int. J. Cancer *34:* 757–761 (1984).
22 Johnson, P.M.; McLaughlin, P.J.; McDicken, I.W.: Serological and tissue localization of placental alkaline phosphatase using monoclonal antibodies in breast and ovarian cancer; in Stigbrand, Fishman, Human alkaline phosphatases, pp. 289–297 (Liss, New York 1984).
23 Johnson, P.M.; Risk, J.M.; Bulmer, J.N.; Niewola, Z.; Kimber, I.: Antigen expression at human materno-fetal interfaces; in Wegmann, Gill, Maternal immunoregulation and fetal survival (Oxford University Press, London 1986).
24 Kellen, J.A.; Bush, R.S.; Malkin, A.: Placental-like alkaline phosphatase in gynecological cancers. Cancer Res. *36:* 269–271 (1976).
25 Loose, J.H.; Damjanov, I.; Harris, H.: Identity of the neoplastic alkaline phosphatase as revealed with monoclonal antibodies to the placental form of the enzyme. Am. J. clin. Path. *82:* 173–177 (1984).
26 Maslow, C.W.; Muensch, H.A.; Azama, F.; Schneider, A.S.: Sensitive fluorometry of heat stable alkaline phosphatase (Regan enzyme) activity in serum of smokers and non-smokers. Clin. Chem. *29:* 260–263 (1983).
27 McDicken, I.W.; Stamp, G.H.; McLaughlin, P.J.; Johnson, P.M.: Expression of human placental-type alkaline phosphatase in primary breast cancer. Int. J. Cancer *32:* 205–209 (1983).
28 McDicken, I.W.; McLaughlin, P.J.; Tromans, P.M.; Luesley, D.M.; Johnson, P.M.: Detection of placental-type alkaline phosphatase in ovarian cancer. Br. J. Cancer *52:* 59–64 (1985).
29 McLaughlin, P.J.; Johnson, P.M.: A search for human placental-type alkaline phosphatases using monoclonal antibodies; in Stigbrand, Fishman, Human alkaline phosphatases, pp. 67–75 (Liss, New York 1984).
30 McLaughlin, P.J.; Gee, H.; Johnson, P.M.: Placental-type alkaline phosphatase in pregnancy and malignancy plasma. Specific estimation using a monoclonal antibody in a solid phase enzyme immunoassay. Clinica chim. Acta *130:* 199–209 (1983).
31 McLaughlin, P.J.; Travers, P.J.; McDicken, I.W.; Johnson, P.M.: Demonstration of placental and placental-like alkaline phosphatases in nonmalignant human tissue extracts using monoclonal antibodies in an enzyme immunoassay. Clinica chim. Acta *137:* 341–348 (1984).
32 McLaughlin, P.J.; Twist, A.M.; Evans, C.C.; Johnson, P.M.: Serum placental-type alkaline phosphatase in cigarette smokers. J. clin. Path. *37:* 826–828 (1984).
33 Millan, J.L.; Stigbrand, T.: Antigenic determinants of human placental and testicular placental-like alkaline phosphatases as mapped by monoclonal antibodies. Eur. J. Biochem. *136:* 1–7 (1983).
34 Millan, J.L.; Eriksson, A.; Stigbrand, T.: A possible new locus of alkaline phosphatase expressed in human testis. Hum. Genet. *62:* 293–295 (1982).
35 Millan, J.L.; Stigbrand, T.; Ruoslahti, E.; Fishman, W.H.: Characterization and use

of an allotype-specific monoclonal antibody to placental alkaline phosphatase in the study of cancer-related phosphatase polymorphism. Cancer Res. *42:* 2444–2449 (1982).

36 Millan, J.L.; Nustad, K.; Norgaard-Pederson, B.: Highly sensitive solid phase immunoenzymometric assay for placental and placental-like alkaline phosphatases with a monoclonal antibody and monodisperse polymer particles. Clin. Chem. *31:* 54–59 (1985).

37 Moss, D.W.: Alkaline phosphatase isoenzymes. Clin. Chem. *28:* 2007–2016 (1982).

38 Mueller, H.D.; Leung, H.; Stimson, R.A.: Different genes code for alkaline phosphatase from human fetal and adult intestine. Biochem. biophys. Res. Commun. *126:* 427–433 (1985).

39 Nakayama, T.; Toshida, M.; Kitamura, M.: *L*-Leucine sensitive heat-stable alkaline phosphatase isoenzyme detected in a patient with pleuritis carcinomatosa. Clinica chim. Acta *30:* 546–548 (1970).

40 Nouwen, E.J.; Pollet, D.E.; Schelstraete, J.B.; Eerdekens, M.W.; Hansch, C.; van de Voorde, A.; De Broe, M.E.: Human placental alkaline phosphatase in benign and malignant ovarian neoplasia. Cancer Res. *45:* 892–902 (1985).

41 Nozawa, S.; Ohta, H.; Arai, H.; Izumi, S.; Kurihara, S.: Heat-stable alkaline phosphatase in normal uterus and deviation of isozyme profile during uterine carcinogenesis; in Stigbrand, Fishman, Human alkaline phosphatases, pp. 223–234 (Liss, New York 1984).

42 Paiva, J.; Damjanov, I.; Lange, P.H.; Harris, H.: Immunohistochemical localization of placental-like alkaline phosphatase in testis and germ cell tumours using monoclonal antibodies. Am. J. Path. *111:* 156–165 (1983).

43 Pollet, D.E.; Nouwen, E.J.; Schelstraete, J.B.; Renard, J.; van de Voorde, A.; De Broe, M.E.: Enzyme-antigen immunoassay for human placental alkaline phosphatase in serum and tissue extracts, and its application as a tumour marker. Clin. Chem. *32:* 41–45 (1985).

44 Robson, E.B.; Harris, H.: Genetics of the alkaline phosphatase polymorphism of the human placenta. Nature, Lond. *207:* 1257–1259 (1965).

45 Sasaki, M.; Fishman, W.H.: Ultrastructural studies on Regan and non-Regan isoenzymes of alkaline phosphatase in human ovarian cancer cells. Cancer Res. *33:* 3008–3018 (1973).

46 Slaughter, C.A.; Coseo, M.C.; Cancro, M.P.; Harris, H.: Detection of enzyme polymorphism by using monoclonal antibodies. Proc. natn. Acad. Sci. USA *78:* 1124–1128 (1981).

47 Slaughter, C.A.; Gogolin, K.J.; Coseo, M.C.; Meyer, L.J.; Lesko, J.; Harris, H.: Discrimination of human placental alkaline phosphatase allelic variants by monoclonal antibodies. Am. J. hum. Genet. *35:* 1–20 (1983).

48 Stigbrand, T.; Engvall, E.: Placental proteins as tumour markers; in Sell, Wahren, Tumor markers (Human Press, Jersey 1982).

49 Sunderland, C.A.; Davies, J.O.; Stirrat, G.M.: Immunohistology of normal and ovarian cancer tissue with a monoclonal antibody to placental alkaline phosphatase. Cancer Res. *44:* 4496–4502 (1984).

50 Swarup, G.; Cohen, S.; Garbers, D.L.: Selective dephosphorylation of proteins containing phosphotyrosine by alkaline phosphatase. J. biol. Chem. *256:* 8197–8201 (1981).

51 Taylor, D.D.; Homesley, H.D.; Doellgast, G.J.: Binding of specific peroxidase-labelled antibody to placental-type phosphatase on tumour-derived membrane fragments. Cancer Res. *40:* 4064–4069 (1980).
52 Tokumitsu, S.; Tokumitsu, K.; Fishman, W.H.: Intracellular alkaline phosphatase activity in cultured human cancer cells. Histochemistry *73:* 1–13 (1981).
53 Tonik, S.E.; Ortmeyer, A.E.; Shindelman, J.E.; Sussman, H.H.: Elevation of serum placental alkaline phosphatase levels in cigarette smokers. Int. J. Cancer *31:* 51–53 (1983).
54 Tucker, D.F.; Oliver, R.I.D.; Travers, P.; Bodmer, W.F.: Serum marker potential of placental alkaline phosphatase-like activity in testicular germ cell tumours evaluated by H17-E2 monoclonal antibody assay. Br. J. Cancer *51:* 631–639 (1985).
55 Uchida, T.; Shimoda, T.; Miyata, H.; Shikata, T.; Iino, S.; Suzuki, H.; Oda, T.; Hirano, K.; Sugiura, M.: Immunoperoxidase study of alkaline phosphatase in testicular tumours. Cancer *48:* 1455–1462 (1981).
56 Van de Voorde, A.; De Groote, G.; De Waele, P.; De Broe, M.E.; Pollet, D.; De Boever, J.; Vandekerckhove, D.; Fiers, W.: Screening of sera and tumor extracts of cancer patients using a monoclonal antibody directed against human placental alkaline phosphatase. Eur. J. Cancer clin. Oncol. *21:* 65–71 (1985).
57 Vockley, J.; D'Souza, M.P.; Foster, C.J.; Harris, H.: Structural analysis of human adult and fetal alkaline phosphatases by cyanogen bromide peptide mapping. Proc. natn. Acad. Sci. USA *81:* 6120–6123 (1984).
58 Williams, G.J.; McLaughlin, P.J.; Johnson, P.M.: Tissue origin of serum placental-like alkaline phosphatase in cigarette smokers. Clinica chim. Acta (in press, 1986).

P.J. McLaughlin, PhD, University Department of Immunology,
Royal Liverpool Hospital, PO Box 147, Liverpool L69 3BX (UK)

Adv. clin. Enzymol., vol. 3, pp. 39–52 (Karger, Basel 1986)

Human Placental Alkaline Phosphatase as a Tumour Marker

M.E. De Broe, E.J. Nouwen, D.E. Pollet, M.W. Eerdekens, T.W. Briers, P.G. Hendrix

Department of Nephrology-Hypertension, University of Antwerp, Wilrijk (Antwerp), Belgium

The discovery in 1968 of an alkaline phosphatase (AP) isoenzyme indistinguishable from hPLAP, in the serum of a patient with bronchogenic carcinoma [21], opened perspectives for the determination of this protein in the serum of cancer patients.

Several methods such as heat inactivation, inhibition, electrophoretic and immunological techniques have been used to identify hPLAP in sera and tissue extracts. All the proposed methods, however, are time consuming and suffer from a lack of specificity and sensitivity. Only the recent introduction of specific monoclonal antibodies (m-Abs) against hPLAP has allowed the development of highly specific, sensitive (0.02 U/l), and reproducible immunoassays and immunohistochemical methods, suited for the determination and/or visualization of minute amounts of this antigen in sera and other biological fluids, and in normal or cancerous tissues [12, 25, 37, 48].

Polymorphism of hPLAP

The heat-stable placental isoenzyme of AP is a highly polymorphic glycoprotein, encoded by a different gene locus than the other two types of AP, i.e. intestinal and tissue-unspecific AP (liver, bone, kidney). Three common alleles give rise to 6 homozygous and heterozygous phenotypes (types 1, 2-1, 2, 3-1, 3-2, and 3), comprising approximately 98% of all the placental variants. The remaining 2% are rare variant enzymes. So far 40 of such rare variants have been identified, based on electrophoretic and immunological differences. The introduction of the hybridoma technology made it possible to study proteins at the level of single antigenic determinants and to look for differences in closely related molecules. As a result, in addition to the discrimination of electrophoretically defined variants of hPLAP, the selective reactivity of m-Abs has already provided clear identification of the non-electrophoretic variants of hPLAP [54].

Trace amounts of a heat-stable hPLAP-like enzyme have been detected in normal human testis [8]. This hPLAP-like enzyme can be distinguished from hPLAP by its different chemical inhibition patterns. Since this testis enzyme was inhibited with *L*-leucine, it was initially thought that normal testis expresses the rare D allele of hPLAP [8]. This concept was, however, severely criticized from a genetic viewpoint. Furthermore, the use of m-Abs indicates structural differences between the testicular hPLAP-like enzyme and hPLAP [41]. Four types of testicular enzymes have been distinguished based on their immunological reactivities, and these differences are compatible with an underlying genetic mechanism. Thus, the testicular hPLAP-like enzyme appears to result from the expression of a locus closely related to but different from the hPLAP locus, which, like the hPLAP locus, would then also be polymorphic [41].

The Regan-type of tumour hPLAP expresses the same allelic variation as is found in term placental AP [26], indicating their identical gene locus origin. The same authors also suggest the identity between another tumour-derived variant of hPLAP, the Nagao isoenzyme, and the rare D-variant of hPLAP, based on similar electrophoretic mobilities and *L*-leucine sensitivity. Recent m-Ab technology, however, indicates that the Nagao enzyme does not represent the reexpression of the D-allele, but appears to reflect the same genetic polymorphism as is found in the placental enzyme [41].

In the past, it was generally thought that human hPLAP is peculiar to the trophoblast, and since it has been shown that the polymorphic hPLAP types are determined by the genotype of the fetus, it was regarded as a purely fetal enzyme [51]. Its expression in certain malignant tumours therefore was considered to be due to an ectopic derepression or reactivation of the fetal AP locus. There is, however, recent evidence for the occurrence of small but significant amounts of heat-stable hPLAP or hPLAP-like enzyme in non-malignant adult tissues (lung, ovary, testis). Therefore, some authors now agree that the occurrence of these enzymes in certain cancer patients is the result of an increased expression and/or release of endogenous hPLAP types by those tissues, following malignant transformation [24, 27, 37].

Serum and Tissue Levels

Polyclonal Antibody Immunoassays

Using polyclonal antisera, elevated levels of hPLAP and hPLAP-like AP have been demonstrated in a number of malignancies, including those from pancreas, lung, breast, colon, lymph nodes, kidney, stomach and bladder [56]. The highest incidence was observed in germ-cell tumours of the testis [27], and ovarian and uterine cancer [7, 52].

Besides their expression in tumour tissue, hPLAP and hPLAP-like AP have also been found in trace amounts in normal tissue. Chang et al. [8] demonstrated the presence of hPLAP-like AP in normal human testis. hPLAP as well as hPLAP-like AP are present in normal endocervix [23], lung and thymus [24].

Monoclonal Antibody Immunoassays

The enzyme-antigen immunoassay used to quantify hPLAP in sera and tissue extracts throughout this study has been described in detail by Pollet et al. [48]. Combining the specificity of m-Abs and the sensitivity of the endogeneous enzyme reaction, a simple, reproducible, and highly sensitive enzyme-antigen immunoassay suited for mass-screening programs was developed. The use of a m-Abs obviates the inaccuracies inherent in inactivation analysis and cross-reactivity of polyclonal antisera.

A first requirement for an effective use of a tumour marker is to establish the normal values in an unselected hospital population. A second requirement is that benign diseases should not interfere with the determination. It was found that 0.1 U/l is the 98th percentile value for hPLAP in an overall hospital population (n = 2,050) without regard to pathology. Using this strategy a benign disease clearly associated with increased hPLAP values was not found [14]. A number of malignant pathologies have been investigated for increased serum hPLAP activity: 48% of ovarian carcinomas were found to be positive, while a population with a variety of malignant pathologies had increased hPLAP values in 11% (table I).

Normal adult ovarian tissue produces almost no hPLAP; it is frequently re-expressed, however, in ovarian neoplasic tissue. This increase in tissue concentration does not always lead to elevated levels of hPLAP in the serum. In a number of ovarian neoplastic tissues, a discrepancy between the *L-p*-bromotetramisole and heat-resistant AP fraction and the hPLAP activity detected by monoclonal antibody immunoassay was found [43]. This

Table I. Frequency of elevated serum hPLAP levels (⩾ 0.1 U/l) in different malignancies: determinations were performed by a monoclonal antibody enzyme-antigen immunoassay

Pathology	n	%
Ovary	68	48
Sarcoma	8	25
Pancreas	10	20
Stomach	26	19
Prostate	6	17
Thyroid	6	17
Colon	22	14
Bladder	14	14
Lung	74	11
Lymphoma	23	9
Breast	100	6
Brain, ileum, larynx, leukaemia, liver, kidney, skin		<5

indicates that only certain types of placental AP are recognized by the m-Ab. A panel of m-Abs directed against different epitopes on hPLAP could result in a higher percentage of detected hPLAP.

Smoking-Associated Alkaline Phosphatase

Since the introduction of hybridoma technology a large number of monoclonal antibodies directed against hPLAP, not cross-reacting with intestinal (IAP) or tissue-unspecific AP have been described by a number of authors [12, 37, 42, 47, 58, 59]. A limited number of these antibodies have been used for the evaluation of hPLAP as a tumour marker in serum and tissue extracts.

There are major differences between these antibodies in their reactivity towards hPLAP-like AP. The antibodies H17E2(59), H7(42) and IQ(Bio)(9) show a pronounced reactivity with hPLAP-like AP, while others like E6(48) and 327 have a rather low reactivity and H317(38) seems to be unreactive. m-Abs reactive with hPLAP-like AP also detect increased hPLAP activity in some heavy smokers. For the effective use of hPLAP as a tumour marker, it is important to avoid interference from smoking habits: this can be achieved by using carefully selected monoclonal antibodies. There are at present no monoclonal antibodies that detect seminoma patients and do not react with hPLAP-like AP. Thus, the need to avoid interference from smoking might mean overlooking seminoma patients.

Immunohistochemical Localisation

Immunohistology is a useful tool for the investigation of several important aspects of a tumour marker, which are difficult by other methods. Some of these questions are: Is hPLAP produced only by malignant cells and by all malignant cells in a tumour and its metastases? What is the subcellular localisation of hPLAP? Is the trace expression of hPLAP in some normal tissues or benign tumours possibly due to local precancerous lesions?

hPLAP in Normal Tissues

There are six sites in the human body where hPLAP can be visualised.

In the normal lung, hPLAP is present on the alveolar plasma membrane of isolated type I pneumocytes of scarce focal areas, and of round-to-cubic respiratory bronchiolar epithelial cells [44].

In the normal adult ovary, hPLAP expression is restricted to the epithelial lining of some germinal inclusion cysts [43, 45], which are occasionally present especially in postmenopausal ovaries. Although not pathological per se, these cystic formations may degenerate into benign or eventually malignant ovarian neoplasms [10]. It is interesting to note that different types of ovarian carcinomas, which are thought to arise from these cysts and from the ovarian germinal epithelium, also have a high hPLAP expression.

Using 5 different m-Abs specific for hPLAP, positive staining has been detected in single intratubular testicular cells, probably of the germ cell series, their number not exceeding 1–2% of all cells [47]. hPLAP staining was also found on the apical plasma membrane of epithelial cells lining the normal oviduct [43, 58] and endocervix and a minor subpopulation of reticular cells in the thymic medulla [58].

The frequent presence of hPLAP in these tissues indicates that in these organs the expression of hPLAP is a normal phenomenon and is not due to precancerous alterations.

Table II. Immunohistology of hPLAP in tumours

Tumour origin	n	n positive	%	Antibody	Reference No.
Ovary	23	13	57	NDOG2	58
	31	10	32	H317	34
	13	12	92	E6	45
	46	27	59	E6, 17E3	unpubl.
Testis	15	11	73	5 m-Abs[1]	47
Breast	7	7	100	H317	35
	8	5	63	5 m-Abs[1]	31
	46	2	4	E6	unpubl.
Colon	4	1	25	5 m-Abs[1]	31
	11	7	64	E6	unpubl.
Rectum	6	2	33	E6	unpubl.
Stomach	20	6	30	E6	unpubl.
Lung	24	6	25	E6	unpubl.
Uterus	11	1	9	E6	unpubl.

[1] Monoclonal antibodies P3/1; SP2/2; SP2/3; SP2/4; SP2/11.

hPLAP in Tumours

Table II summarizes some results of the available immunohistochemical studies on the expression of hPLAP in tumour tissues. From these reports it is evident that the histological distribution of hPLAP in malignant tumours is, in general, heterogenous; i.e. a substantial number of malignant cells is not stained. In addition, hPLAP is always predominantly localized on the apical or the entire plasma membrane of malignant cells. Male and female germ cell tumours have a much higher frequency and extent of hPLAP immunohistochemical staining than all other neoplasms. This may be due to the fact that the fetal gonads are, besides the chorionic villi, the only fetal organs or tissues which express hPLAP.

In malignant ovarian tumours, hPLAP staining is most frequent (75–85%) and most abundant in papillary serous cystadenocarcinomas, and somewhat less frequent in endometrioid carcinomas, undifferentiated carcinomas and granulosa cell tumours [34, 45, 58]. Likewise, in the group of benign tumours, serous cystadenomas and cystadenofibromas have the highest incidence (100%). We could not confirm the reported presence of cytoplasmic staining for hPLAP in some mucinous tumours [58].

Among the testicular neoplasms, seminomas have the highest frequency of hPLAP-positive staining (86–100%). In non-seminomas, hPLAP immunostaining is restricted to seminoma, embryonic carcinoma, and choriocarcinoma components. The use of five different monoclonal antibodies suggests that in some instances the tumoral isoenzyme is antigenically different from hPLAP found in fetal or adult testicular germ cells [47].

Although 60% of colonic adenocarcinomas are positively stained for hPLAP [43], staining intensity is in general less important than in ovarian or testicular neoplasms.

hPLAP in Fetal Tissues

In full-term placenta, intense hPLAP staining is localized on the apical and less-intense staining on the basal plasma membrane of syncythiotrophoblastic cells [45, 47]. In the 11-week placenta, only some villi are positive. The number of positively stained villi increases with gestation time.

In the fetal ovary of 13–14 weeks of gestation, hPLAP staining is found in a substantial number of cells, which increase in number towards the periphery. They probably are germ cells [Nouwen et al., unpubl.]. The fetal ovarian surface epithelium is negative.

In the fetal testis staining for hPLAP is observed in some intratubular cells [47].

Comparison of hPLAP with Other Tumour-Associated Antigens in Ovarian Cancer (table III)

Immunohistology

Nearly all borderline and malignant nonmucinous tumours are positive for CA 125 [29, 43]. Normal organs expressing CA 125 are the ovary, pleural and peritoneal mesothelium, lung, breast, fallopian tube, and endocervix [28, 43, 44, Nouwen et al., unpubl.].

Ninety percent of mucinous adenocarcinomas are positive for CA 19-9, but only 21% of serous adenocarcinomas [29]. A substantial number of normal organs also produce CA 19-9 in small amounts, i.e. stomach, colon, pancreas, salivary gland, liver, gall bladder, bronchial glands, prostate, and endocervix [2].

Table III. hPLAP and CA 125 in body fluids and ovarian tissue

Sample		n	PLAP, % positive	n	CA 125 % positive
Serum	healthy subjects	150	0	888[1]	1
	patients with liver disease	47	0	47	45
	smokers	30	0[5]		ND
	ovarian cancer patients	56	48	101[1]	83
Ascites	benign ovarian tumours	6	0	8	100
	malignant ovarian tumours	8	88	8	100
Tissue extracts	normal ovaries	14	0		ND
	benign ovarian tumours	17	0		ND
	malignant ovarian tumours	36	58		ND
Immuno-histology	normal ovaries	23	4[2]	23	65[3]
	benign ovarian tumours	27	22[4]	27	63
	malignant ovarian tumours	51	61	51	82

[1] Bast et al. [3].
[2] Some germinal inclusion cysts are positive; all other constituents are negative.
[3] Inclusion cysts, germinal epithelium, invaginations and little proliferations of this epithelium.
[4] Some serous cysts and cystadeno(fribro)mas.
[5] Heavy smoking may result in false positives.

Only 21% of serous and 10% of mucinous adenocarcinomas are positive for CEA [29]. Normal CEA-expressing organs are the lung, stomach, colon, and rectum.

Serum

Increased serum levels of CA 125 are found in 50–92% of ovarian cancer patients [3, 14, 50], compared to 48% for hPLAP [14, 45]. Increased levels of CA 19-9 are found in 25–29% [50] and CEA in 4–10% of these patients [14, 50].

A serious disadvantage for CA 125 as a tumour marker is its high frequency of false-positivity (45%) in chronic liver disease [14]. CA 19-9 also has a relatively high rate of false-positivity (27%) in benign liver pathologies [13]. hPLAP is not increased in sera of patients affected by these benign pathologies, except in some heavy smokers [14].

Table IV. Tumour imaging by immunoscintigraphy with radiolabelled monoclonal antibodies in cancer patients

Carcinoma	m-Ab	Antigen	Success rate	Isotope	Reference No.
Ovary, breast	HMFG1 and 2	HMFG	15/20	^{123}I	18
Colorectal	791T/36	colorectal	10/11	^{131}I	20
Gastrointestinal	anti-CEA	CEA	16/17	^{131}I	4
Melanoma	anti-P8.2	P97 protein	9/9	^{131}I	30
Colorectal	17-1A	colorectal	34/63	^{131}I	32
Breast	M8	HMFG	22/29	^{111}In	48
Colon, breast, teratoma	YPC2/12.1	colorectal	16/27	^{131}I	54
Gastrointestinal	791T/36	Sarcoma	28/54	^{131}I	1
Lymphoma	T101	lymphoma	1/1	^{111}In	6
Rectal	AUAI		1/1	^{111}In	19
Testis, ovary, cervix	H17E2	hPLAP	9/15	^{111}In	16
Ovary	H317	hPLAP	11/18	N	33
Ovary	$NDOG_2$	hPLAP	16/20	^{123}I	11

Perspectives and Prospectives of hPLAP

There is already a lot of information indicating that m-Abs to tumour-associated antigens will eventually have clinically applications in tumour imaging and therapy. The main reason for this is the possibility of scaling therapy and imaging down to the selected tumour cells and thus leaving the normal cells of the patient unaffected [5, 53].

The usefulness of m-Abs to tumour markers for diagnostic imaging has already been demonstrated in several studies (table IV). Epenetos et al. [16] showed that m-Abs to hPLAP and to testicular hPLAP-like AP (H17E2) labelled with indium-111 and used in immunoscintigraphy gave good images of neoplastic lesions in most patients with active disease. For some patients, the technique even provided information which was not obtainable with other imaging techniques, including ultrasonography and CT. With another radiolabelled anti-hPLAP m-Ab, H317, patients suspected of ovarian cancer recurrence showed focal increased abdominal uptake of the radiolabelled m-Abs [36]. These two studies clearly show the potential of hPLAP as a tumour marker for imaging.

The potential of hPLAP in drug targeting is demonstrated by Tsukazaki et al. [60]. They demonstrated that ricin A conjugates of anti-α-fetoprotein show no specific cytotoxity to α-fetoprotein-producing hepatoma cells, while similar conjugates of anti-hPLAP specifically kill tumour cells producing hPLAP. Their results suggest that the difference is due to α-fetoprotein being a secreted protein, and hPLAP, a membrane protein. Because hPLAP is a membrane protein, it is far superior to α-fetoprotein for therapy or imaging. These and other studies [15, 17, 33, 39, 62] demonstrate the promising diagnostic and therapeutic value of hPLAP, and of tumour-associated antigens in general.

Acknowledgments

We wish to thank Dr. André Van de Voorde (Department of Molecular Biology, State University of Ghent, Belgium) for providing us the monoclonal antibody E6, Dr. Robert C. Bast (Department of Medicine, Duke University Medical Center, Durham, N.C.) for the batch of antibody OC 125, and Drs. P.J. Den Houwdyker, J. Renard, L. Delbeke, P. Buytaert, F. Uytenbroeck, I. De Wever, and J. Claeys for providing some of the biopsies. The technical assistance of S. Dauwe, E. Martens, and R. Marynissen and the secretarial skills of E. Snelders are once more greatly acknowledged.

References

1 Armitage, N.C.; Perkins, A.C.; Pimm, M.V.; Farrands, P.A.; Baldwin, R.W.; Hardcastle, J.D.: The localization of an anti-tumour monoclonal antibody (791T/36) in gastrointestinal tumours. Br. J. Surg. *71:* 407–412 (1984).

2 Atkinson, B.F.; Ernst, C.; Herlyn, M.; Steplewski, Z.; Sears, H.; Koprowski, H.: Gastrointestinal cancer-associated antigen in immunoperoxidase assay. Cancer Res. *42:* 4820–4823 (1982).

3 Bast, R.; Klug, T.; John, E.; Niloff, J.; Lazarus, H.; Berkowitz, R.;. Leavitt, T.; Griffiths, T.; Parker, L.; Zurawski, V.; Knapp, R.: A radioimmunoassay using a monoclonal antibody to monitor the course of epithelial ovarian cancer. New Engl. J. Med. *309:* 883–887 (1983).

4 Berche, C.; Mach, J.P.; Lumbrosco, J.D.; Langlais, C.; Aubry, F.; Buchegger, F.; Carrel, S.; Rougier, P.; Parmentier, C.; Tubiana, M.: Tomoscintigraphy for detecting gastrointestinal and medullar thyroid cancers. First clinical results using radiolabelled monoclonal antibodies against carcinoembryonic antigen. Br. med. J. *285:* 1447–1451 (1982).

5 Bloomer, W.D.; Lipsztein, M.; Dalton, J.F.: Antibody-mediated radiotherapy. Cancer *55:* 2229–2233 (1985).

6 Bunn, P.A.; Carrasquillo, J.A.; Keenan, A.M.; Schroff, R.W.; Foon, K.A.; Hsu, S.M.;

Gazdar, A.F.; Reynolds, J.C.; Parentesis, P.; Larson, S.: Imaging of T-cell lymphoma by radiolabelled monoclonal antibody. Lancet *ii:* 1219–1221 (1984).
7 Cadeau, B.J.; Blackstein, M.E.; Malkin, A.: Increased incidence of placental-like alkaline phosphatase activity in breast and genitourinary cancer. Cancer Res. *34:* 729–732 (1974).
8 Chang, C.H.; Angellis, D.; Fishman, W.H.: Presence of the rare D-variant heat-stable, placental-type alkaline phosphatase in normal human testis. Cancer Res. *40:* 1506–1510 (1980).
9 Cooper, E.H.; Pidcock, N.B.; Jones, W.G.; Ward, A.M.: Evaluation of an amplified enzyme-linked immunoassay of placental alkaline phosphatase in testicular cancer. Eur. J. Cancer clin. Oncol. *21:* 525–530 (1985).
10 Czernobilsky, B.: Primary epithelial tumors of the ovary; in Blaustein, Pathology of the female genital tract, pp. 453–504 (Springer, Berlin 1977).
11 Davies, E.R.; Davies, J.O.; Jackson, P.C.; Howe, K.; Pitcher, E.M.; Sadowski, C.S.; Stirrat, G.M.; Sunderland, C.A.: ^{123}I-NDOG$_2$ anti-placental alkaline phosphatase imaging of ovarian tumours. Br. J. Cancer *52:* 649 (1985).
12 De Groote, G.; De Waele, P.; Van de Voorde, A.; De Broe, M.; Fiers, W.: Use of monoclonal antibodies to detect human placental alkaline phosphatase. Clin. Chem. *29:* 115–119 (1983).
13 Delvillano, B.C.; Zurawski, V.: The carbohydrate antigenic determinant 19-9 (CA 19-9): a monoclonal antibody defined tumor marker. Immunodiagnostics, pp. 269–282 (Liss, New York 1983).
14 Eerdekens, M.W.; Nouwen, E.J.; Pollet, D.E.; Briers, T.W.; De Broe, M.E.: Placental alkaline phosphatase and cancer antigen 125 in sera of patients with benign and malignant diseases. Clin. Chem. *31:* 687–690 (1985).
15 Engelstad, B.L.; Wesley, G.E.; Huberty, J.P.; Moseley, M.E.; McNamara, M.T.; Triglia, R.P.; Stitla, L.A.; Brasch, R.C.; Hattner, R.S.: In vitro and in vivo testing of metal-ion complexes used as nuclear magnetic resonance contrast media. Investve Radiol. *19:* 149–150 (1984).
16 Epenetos, A.A.; Hooker, G.; Durbin, H.; Bodmer, W.F.; Snook, D.; Begent, R.; Oliver, R.T.D.; Lavender, J.P.: Indium-111 labelled monoclonal antibody to placental alkaline phosphatase in the detection of neoplasms of testis, ovary, and cervix. Lancet *ii:* 350–353 (1985).
17 Epenetos, A.A.: Clinical results with regional antibody-guided irradiation. Br. J. Cancer *52:* 653 (1985).
18 Epenetos, A.A.; Mather, S.; Granowska, M.; Nimmon, C.C.; Hawkins, L.R.; Britton, R.E.; Shepperd, J.; Taylor-Papadimitriou, J.; Durbin, H.; Malpar, J.S.: Targeting of iodine-123 labelled monoclonal antibodies to ovarian, breast, and gastrointestinal tumours. Lancet *ii:* 999–1004 (1982).
19 Epenetos, A.A.; Snook, D.; Hooker, G.; Lavender, J.P.; Halman, K.E.: Tumour imaging using an improved method of DTPA-coupled monoclonal antibodies radiolabelled with metallic radionuclides. Lancet *ii:* 169 (1984).
20 Farrands, P.A.; Pimm, M.V.; Embleton, M.J.; Hardcastle, J.D.; Perkins, A.C.; Hardy, J.D.; Baldwin, R.W.: Radioimmunodetection of human colorectal cancers by an anti-tumour monoclonal antibody. Lancet *ii:* 397–400 (1982).
21 Fishman, W.H.; Inglis, N.R.; Stolbach, L.L.; Krant, M.J.: Serum alkaline phosphatase isoenzyme of human neoplastic origin. Cancer Res. *28:* 150–154 (1968).

22 Fishman, W.H.; Inglis, N.R.; Vaitukaitis, J.; Stolbach, L.L.: Regan isoenzyme and human chorionic gonadotropin in ovarian cancer. Natn. Cancer Inst. Monogr. *42:* 63–73 (1975).
23 Goldstein, D.J.; Blasco, L.; Harris, H.: Placental alkaline phosphatase in nonmalignant human cervix. Proc. natn. Acad. Sci. USA *77:* 4226–4228 (1980).
24 Goldstein, D.J.; Rogers, L.; Harris, H.: A search for trace expression of placental-like alkaline phosphatase in non-malignant human tissues. Demonstration of its occurrence in lung, cervix and thymus. Clinica chim. Acta *125:* 63–75 (1982).
25 Haije, W.G.; Meerwaldt, J.H.; Talerman, A.; Kuipers, T.J.; Baggerman, L.; Teeuw, A.H.; Van der Pompe, W.B.; Van Driel, J.: The value of a sensitive assay of carcinoplacental alkaline phosphatase (CPAP) in the follow-up of gynecological cancers. Int. J. Cancer *24:* 288–293 (1979).
26 Inglis, N.R.; Kirley, S.; Stolbach, L.L.; Fishman, W.H.: Phenotypes of the Regan Isoenzyme and identity between the placental D-variant and the Nagao isoenzyme. Cancer Res. *33:* 1657–1664 (1973).
27 Jeppson, A.; Wahren, B.; Brehmer-Andersson, E.; Silfversward, C.; Stigbrand, T.; Millan, J.L.: Eutopic expression of placental-like alkaline phosphatase in testicular tumors. Int. J. Cancer *34:* 757–761 (1984).
28 Kabawat, S.E.; Bast, R.; Bhan, A.; Welch, W.; Knapp, R.; Colvin, R.: Tissue distribution of a coelomic epithelium-related antigen recognized by the monoclonal antibody OC 125. Lab. Invest. *48:* 42A (1983).
29 Kremer, M.; Chatal, J.; Curtet, C.; Douillard, J.: The immunohistochemical reactivity of monoclonal antibodies with tissue sections of human ovarian tumors. Proc. XIth Ann. Meet. Int. Soc. Oncodevelopmental Biology and Medicine, Stockholm 1983.
30 Larson, S.M.; Brown, J.P.; Wright, P.W.; Carrasquillo, J.A.; Hellström, I.; Hellström, K.E.: Imaging of melanoma with I-131-labelled monoclonal antibodies. J. nucl. Med. *24:* 123–129 (1983).
31 Loose, J.H.; Damjanov, I.; Harris, H.: Identity of the neoplastic alkaline phosphatase as revealed with monoclonal antibodies to the placental form of the enzyme. Am. J. clin. Path. *82:* 173–177 (1984).
32 Mach, J.P.; Chatal, J.F.; Lumbroso, J.D.; Buchegger, F.; Forni, M.; Ritschard, J.; Berche, C.; Douillard, J.Y.; Carrel, S.; Herlyn, M.: Tumor localization in patients by radiolabeled monoclonal antibodies against colon carcinoma. Cancer Res. *43:* 5593–5600 (1983).
33 Marx, J.L.: Antibodies made to order. Science *229:* 455–456 (1985).
34 McDicken, I.W.; McLaughlin, P.J.; Tromans, P.M.; Luesley, D.M.; Johnson, P.M.: Detection of placental-type alkaline phosphatase in ovarian cancer. Br. J. Cancer *52:* 59–64 (1985).
35 McDicken, I.W.; Stamp, G.H.; McLaughlin, P.J.; Johnson, P.M.: Expression of human placental-type alkaline phosphatase in primary breast cancer. Int. J. Cancer *32:* 205–209 (1983).
36 McLaughlin, P.J.; Critchley, M.; Tromans, P.M.; McDicken, I.W.; Johnson, P.M.: Placental-type alkaline phosphatase detection by monoclonal antibody (H317). Application in ovarian cancer recurrence. Br. J. Cancer *52:* 667 (1985).
37 McLaughlin, P.J.; Travers, P.J.; McDicken, I.W.; Johnson, P.M.: Demonstration of placental and placental-like alkaline phosphatase in non-malignant human tissue

extracts using monoclonal antibodies in an enzyme immunoassay. Clinica chim. Acta *137:* 341–348 (1984).

38 McLaughlin, P.J.; Twist, A.M.; Evans, C.C.; Johnson, P.M.: Serum placental type alkaline phosphatase in cigarette smokers. J. clin. Path. *37:* 826–828 (1984).

39 Mew, D.; Wat, C.K.; Towers, G.H.N.; Levy, J.G.: Ability of specific monoclonal antibodies and conventional antisera conjugated to hematoporphyrin to label and kill selected cell lines subsequent to light activation. Cancer Res. *45:* 4380–4386 (1985).

40 Millan, J.L.; Beckman, G.; Jeppson, A.; Stigbrand, T.: Genetic variants of placental alkaline phosphatase as detected by a monoclonal antibody. Hum. Genet. *60:* 145–149 (1982).

41 Millan, J.L.; Stigbrand, T.: Antigenic determinants of human placental and testicular placental-like alkaline phosphatases as mapped by monoclonal antibodies. Eur. J. Biochem. *136:* 1–7 (1983).

42 Millan, J.L.; Stigbrand, T.; Ruoslahti, E.; Fishman, W.H.: Characterization and use of an allotype-specific monoclonal antibody to placental alkaline phosphatase in the study of cancer-related phosphatase polymorphism. Cancer Res. *42:* 2444–2449 (1982).

43 Nouwen, E.J.; Eerdekens, M.W.; Briers, T.W.; Pollet, D.E.; De Broe, M.E.: Immunohistochemical localisation of hPLAP, CA 125 and CEA in normal and neoplastic human tissues and the foetus. Br. J. Cancer *52:* 644–645 (1985).

44 Nouwen, E.J.; Pollet, D.E.; Eerdekens, M.W.; Hendrix, P.G.; Briers, T.W.; De Broe, M.E.: Immunohistochemical localization of placental alkaline phosphatase, carcinoembryonic antigen, and cancer antigen 125 in normal and neoplastic human lung. Cancer Res. *46:* 866–876 (1986).

45 Nouwen, E.J.; Pollet, D.E.; Schelstraete, J.B.; Eerdekens, M.W.; Hänsch, C.; Van de Voorde, A.; De Broe, M.E.: Human placental alkaline phosphatase in benign and malignant ovarian neoplasia. Cancer Res. *45:* 892–902 (1985).

46 Nozawa, S.; Ohta, M.; Izumi, S.; Kayashi, S.; Tsutsni, F.; Kurihara, S.; Watanabe, K.: Heat-stable alkaline phosphatase in the normal female genital organ – with special reference to the histochemical heat-stability test and L-phenylalanine inhibition test. Acta histochem. cytochem. *13:* 521–530 (1980).

47 Paiva, J.; Damjanov, I.; Lange, P.; Harris, H.: Immunohistochemical localization of placental-like alkaline phosphatase in testis and germ-cell tumors using monoclonal antibodies. Am. J. Path. *111:* 156–165 (1983).

48 Pollet, D.E.; Nouwen, E.J.; Schelstraete, J.B.; Renard, J.; Van de Voorde, A.; De Broe, M.E.: Enzyme-antigen immunoassay for human placental alkaline phosphatase in serum and tissue extracts, and its application as a tumor marker. Clin. Chem. *31:* 41–45 (1985).

49 Rainsbury, R.M.; Ott, R.J.; Westwood, J.H.; Kalirai, T.S.; Coombes, R.C.; McCready, V.R.; Neville, A.M.; Gazet, J.C.: Localization of metastatic breast carcinoma by a monoclonal antibody chelate labelled with indium-111. Lancet *ii:* 934–938 (1983).

50 Ricolleau, G.; Chatal, J.-F.; Fumoleau, P.; Kremer, M.; Douillard, J.-Y.; Curtet, C.: Radioimmunoassay of the CA 12-5 antigen in ovarian carcinomas. Advantages compared with CA 19-9 and CEA. Tumour Biol. *5:* 151–159 (1984).

51 Robson, E.B.; Harris, H.: Genetics of the alkaline phosphatase polymorphism of the human placenta. Nature, Lond. *207:* 1257–1259 (1965).

52 Sasaki, M.; Fishman, W.H.: Ultrastructural studies on Regan and non-Regan isoenzymes of alkaline phosphatase in human ovarian cancer cells. Cancer Res. *33:* 3008–3018 (1973).
53 Sikora, K.; Smedley, H.; Thorpe, P.: Tumour imaging and drug targeting. Br. med. Bull. *40:* 233–239 (1984).
54 Slaughter, G.A.; Coseo, M.; Cangro, M.; Harris, H.: Detection of enzyme polymorphism by using monoclonal antibodies. Proc. natn. Acad. Sci. USA *78:* 1124–1128 (1981).
55 Smedley, H.M.; Finan, P.; Lennox, E.S.; Ritson, A.; Takei, F.; Wraight, P.; Sikora, K.: Localisation of metastatic carcinoma by a radiolabelled monoclonal antibody. Br. J. Cancer *47:* 253–259 (1983).
56 Stigbrand, T.; Engvall, E.: Placental proteins as tumor markers; in Sell, Wahren, Tumor markers (Humana Press, New Jersey 1982).
57 Stigbrand, T.; Holmgren, P.A.; Jeppson, A.; Damber, M.-G.; Schoultz, B. von: On the value of placental alkaline phosphatase as a marker for gynecological malignancy. Acta obstet. gynec. scand. *64:* 99–103 (1985).
58 Sunderland, C.; Davies, J.; Stirrat, G.: Immunohistology of normal and ovarian cancer tissue with a monoclonal antibody to placental alkaline phosphatase. Cancer Res. *44:* 4496–4502 (1984).
59 Travers, P.J.; Bodmer, W.F.: Preparation and characterisation of monoclonal antibodies against placental alkaline phosphatase and other human trophoblast associated determinants. Int. J. Cancer *33:* 633–641 (1984).
60 Tsukazaki, K.; Hayman, E.G.; Ruoslahti, E.: Effects of Ricin A chain conjugates of monoclonal antibodies to human alfa-fetoprotein and placental alkaline phosphatase on antigen-producing tumor cells in culture. Cancer Res. *45:* 1834–1838 (1985).
61 Wolf, G.L.; Burnett, K.R.; Goldstein, E.J.; Joseph, P.M.: Contrast agents for magnetic resonance imaging; in Kressel, Magnetic resonance annual, pp. 231–266 (Raven Press, New York 1985).

M.E. De Broe, MD, PhD, Nephrology-Hypertension Unit,
c/o University of Antwerp – A.Z., Wilrijkstraat 10, B–2520 Edegem (Belgium)

Adv. clin. Enzymol., vol. 3, pp. 53–59 (Karger, Basel 1986)

Radiolabelled Monoclonal Antibodies to Placental Alkaline Phosphatase in the Diagnosis and Treatment of Neoplasms of the Testis, Ovary and Cervix

T.R.B. Pawlikowska, A.A. Epenetos

on behalf of the Hammersmith Oncology Group and
the Imperial Research Fund, London[1]

Radiolabelled monoclonal antibodies have been demonstrated to localize preferentially to neoplastic tissue [1]. However, cross-reaction with normal tissues and in vivo dehalogenation are significant limitations. These may be minimized by careful selection of monoclonal antibody to maximize the specificity and, furthermore, targetting may be predicted by prior assessment of antibody reactivity with tissue sections of neoplastic and normal tissue. Alteration of the specific radiolabel can avoid dehalogenation.

Hence, to achieve better localization, monoclonal antibodies with a restricted specificity are required (i.e. low reactivity with normal tissue, whilst having a high reactivity with neoplastic tissue). As monoclonal antibodies will only react avidly with a single antigenic epitope, it is desirable

[1] The following people from the Royal Postgraduate Medical School, Hammersmith Hospital, Charing Cross Hospital, Institute of Urology and the Imperial Cancer Research Fund, London have been involved in this study: R. Begent, W.F. Bodmer, J. Burchell, C. Coulter, A. Cross, N. Courtenay-Luck, H. Durbin, B. Dhokia, K.E. Halnan, B. Henderson, T. Krausz, G. Hooker, H.E. Lambert, J.P. Lavender, W. MacGregor, C.J. McKenzie, R. Rampling, A.J. Munro, N. McWhinney, M. Myers, J.S. Orr, D. Pectasides, D. Snook, A. Rahemtulla, J. Shepherd, M. Slevin, A. Stewart Ross, R.D.T. Oliver, J. Taylor-Papadimitriou, S. Van Noorden and J. Walford.

The following from the University of Liverpool have also been involved in this study: P.M. Johnson and J. McLaughlin.

that the antigen is present in high concentrations on tumour cells to optimize targetting. Following a consideration of the above, two monoclonal antibodies (H317 and H17-E2) against placental alkaline phosphatase (PLAP) and testicular placental-like alkaline phosphatase were selected to be evaluated as radioimmunolocalizing agents. In addition, a new indium-111 (^{111}In) radiolabelling method was used as it has been shown both in experimental models and clinical trials to produce better tumour to normal organ ratios than radioactive iodine [6].

PLAP is expressed on the surface membrane of many epithelial ovarian cancers and testicular germ cell tumours [2, 5]. The amount of PLAP in ovarian tissue samples has been shown to have a good correlation with elevated serum PLAP levels noted in patients. Serum PLAP assays in patients with testicular germ cell tumours have shown consistent elevation only in patients with active seminomatous neoplasms [7].

Immunohistochemically, no positive staining of PLAP on cells of any normal fresh frozen tissue has been found (excepting term placenta). The antigen is expressed at trace levels in some normal tissues [8]. PLAP is a potentially useful tumour-associated antigen in experimental models. Monoclonal antibodies against PLAP have produced good immunolocalization of PLAP-producing tumours [3].

Patients and Methods

Patients with carcinoma of the ovary, cervix and testis formed part of the scanning study. Patients with relapsed advanced carcinoma of the ovary were considered for monoclonal antibody guided therapy.

Eighteen patients with carcinoma of the ovary, 20 patients with carcinoma of the testis and 1 patient with carcinoma of the cervix, were scanned with H17-E2 antibody coupled to iodine-123 or indium-111 [6, 12].

In addition, 12 patients with metastatic carcinoma of the ovary (FIGO stages III and IV) aged 36–75 years (mean 58 years), who had relapsed following surgery, chemotherapy and/or radiotherapy were considered for monoclonal antibody guided treatment. Anti-PLAP monoclonal antibody was used, together with a panel of other monoclonals, as deemed appropriate, following immunohistochemical studies on each patient's tumour tissue.

Prior to therapy, any ascites was drained and 1 mCi of antibody was injected intraperitoneally and washed in with normal saline. Scans were obtained for quantitation and dosimetry prior to administering a therapy dose of 20–150 mCi via this route.

Written and informed consent was obtained prior to entry into the study. Patients were skin tested for allergy to mouse immunoglobulin prior to injection and were given potassium iodide 120 mg/day for 7 days, starting 1 day before the injection.

Monoclonal Antibodies

H17-E2. This is a mouse IgG_1 and was raised against purified plasma membranes of normal term placenta. It precipitates placental alkaline phosphatase activity as a single band of 67,000 daltons consistent with the molecular weight of PLAP. In addition, it reacts with the leucine-inhibitable form of alkaline phosphatase found at low levels in the normal testis and which cross-reacts with the placental enzyme. It does not react with the non-placental forms of alkaline phosphatase [5].

Immunohistochemical testing has shown it to react strongly with a wide range of neoplasms, but especially those of the testis, ovary and cervix [2]. It has also been shown to react negatively against normal tissue, excepting normal term placenta. However, PLAP has been found to be present at trace levels in some normal tissues [8].

H317. This is also a mouse IgG_1 which was produced by immunizing with syncytiotrophoblast microvilli from preparations of term placentae. It is specific for the heat stable *L*-phenylalanine-inhibitable PLAP. This antibody does not react with the testicular variant of PLAP and other non-placental forms of alkaline phosphatase [8].

Large-scale antibody production, purification and immunotesting were performed as previously described. Sterility and pyrogenicity testing was conducted by an independent laboratory (Safepharm Labs).

Radiolabelling

Antibodies were labelled with iodine-123 (AERE Harwell) or iodine-131 (Amersham International) using the iodogen method [11]. Anti-PLAP antibody was also labelled with ^{111}In (INSI Amersham International) using the DTPA anhydride method [6]. Tests for aggregate formation, stability and immunoreactivity were satisfactory for both systems.

Imaging

Gamma-camera scans were obtained immediately after injection (blood pool image) and sequentially for up to 5 days with iodine labelled monoclonal antibody and 7 days for indium-labelled antibody. Emission tomography (SPECT) and whole body scans were performed if indicated. Some patients ($n = 6$) received monoclonal antibody on more than one occasion (maximum 5 times).

Results

Figure 1 shows a typical example of an indium scan, using H17-E2 monoclonal antibody in a patient with a relapsed testicular tumour. There is a region of increased uptake in the left para-aortic area (indicated by the arrow) which was reported as being abnormal on the basis of this scan. This patient had a rise in alphafeto-protein tumour marker levels, despite an initial response to chemotherapy, and underwent restaging.

The results of all investigations (including a CT scan of the abdomen) were normal except for the monoclonal antibody scan. A laparotomy was

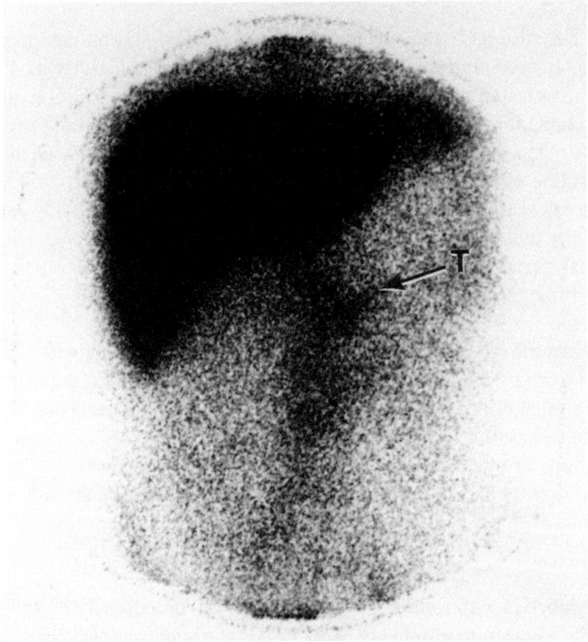

Fig. 1. Image at 48 h showing uptake in para-aortic lymph node as marked.

therefore performed. Despite normal appearances, histology of a biopsy specimen taken from the para-aortic nodes near the hilum of the left kidney proved to be positive for metastatic teratoma. The monoclonal antibody scan had thus detected active disease, not found by conventional imaging.

As detailed elsewhere [9], of 15 patients scanned with indium-labelled monoclonal antibody H17-E2, positive scans were obtained in 10 patients, compared with negative scans in 5 patients. There was good correlation with scan results and histology. In each case, the findings of antibody imaging were compared with those obtained from conventional investigations. No false-positive scans were obtained and no false-negative scans were seen in patients with PLAP-positive tumours. Images were good enough not to need further computer manipulation. It was found that images were optimal at 2 days or later, after the time of injection. However, in all patients, there was high liver uptake of indium and therefore, this radiolabel is unsuitable for imaging hepatic metastases.

In the study of 12 patients with advanced carcinoma of the ovary, it was concluded that most patients had benefited symptomatically. Those with stage III carcinoma of the ovary and minimal residual disease (less than 2 cm diameter) were found to respond optimally. Four out of six patients in this group achieved a complete remission (longest follow-up 2 years), having failed to respond to both chemotherapy and surgery. Hence, the results obtained so far compare favourably with intraperitoneal chemotherapy and need to be investigated further with a larger study.

In treating patients with advanced relapsed carcinoma of the ovary, it was concluded that patients without ascites, who had not been exposed to prior radiotherapy, may tolerate 100–151 mCi of iodine-131 monoclonal antibody intraperitoneally. Side effects are leucopenia, thrombocytopenia and diarrhoea (all effects were reversible).

This dose threshold must be lowered to not more than 100 mCi in patients with ascites but no prior radiotherapy, and patients with ascites who have had previous radiotherapy. Hence, radiotherapy toxicity is avoided.

Discussion

This study has shown that radiolabelled monoclonal antibodies to PLAP can successfully image tumour masses in patients with PLAP-positive tumours, i.e. testicular, ovarian and cervical cancers. It has been shown immunohistochemically that radiolabelled PLAP monoclonal antibodies produce good tumour-to-normal-tissue ratios in these neoplasms. In a few patients, this technique yielded information with regard to the site of active tumour unavailable through other radiological imaging techniques [9].

The quality of images obtained with ^{123}I was preferable to those seen with ^{131}I. However, dehalogenation of labelled antibody remains a problem, since the free ^{123}I is available for uptake by normal thyroid and stomach. Good quality images were obtained with ^{111}In-labelled antibody which was used to try to circumvent this. Unfortunately, ^{111}In is taken up by the liver and the intense image thus produced renders antibody labelled with this isotope unsuitable for visualizing hepatic metastases. As an alternative to the above, iodinated monoclonal antibody could be introduced intraperitoneally into patients with, for example, cancer of the ovary, to detect intra-abdominal disease. In this way, dehalogenation could be minimized [12, 13].

Positive monoclonal antibody scans were not obtained in all patients with active neoplasms. It is thought that this may be accounted for by variable expression of PLAP, both between patients and tumour (with the proviso that high serum PLAP levels have been noted in some cigarette smokers).

In conclusion, radiolabelled monoclonal antibodies to PLAP have been shown to successfully target sites of active neoplasia. This method of scanning should be considered for diagnosis, staging and monitoring of patients with tumours of the testis, ovary and cervix. A significant number of patients are eminently curable, especially if detection occurs early in their disease.

Some therapeutic benefit has ensued from using radiolabelled anti-PLAP monoclonal antibody in treating advanced carcinoma of the ovary. However, further trials are needed to evaluate this approach fully.

References

1 Epenetos, A.A.; Britton, K.E.; Mather, S.: Targeting of Iodine-123-labelled tumour-associated monoclonal antibodies to ovarian, breast and gastrointestinal tumours. Lancet *ii:* 999–1005 (1982).

2 Epenetos, A.A.; Travers, P.; Gatter, K.: An immunohistological study of testicular germ cell tumours using two different monoclonal antibodies against placental alkaline phosphatase. Br. J. Cancer *49:* 11–15 (1984).

3 Epenetos, A.A.: Monoclonal antibodies in the localisation of human neoplasms in vitro and in vivo; PhD thesis, University of London (1983).

4 Arklie, J.; Taylor-Papadimitriou, J.; Bodmer, W.F.; Egan, M.; Millis, R.: Differentiation antigens expressed by epithelial cells in the lactating breast are also detectable in breast cancers. Int. J. Cancer *28:* 23–29 (1981).

5 Travers, P.; Bodmer, W.F.: Preparation and characterisation of monoclonal antibodies against placental alkaline phosphatase and other human trophoblast associated determinants. Int. J. Cancer *33:* 633–641 (1984).

6 Hnatowich, D.J.; Layne, W.W.; Childs, R.L.: Radioactive labelling of antibody. A simple and efficient method. Science *220:* 613–615 (1983).

7 Epenetos, A.A.; Munro, A.J.; Tucker, D.; Gregory, W.; Duncan, W.; MacDougall, R.H.; Faux, M.; Travers, P.; Bodmer, W.E.: Serum placental alkaline phosphatase assays in the monitoring of testicular tumours. Br. J. Cancer *51:* 641–644 (1985).

8 McLaughlin, P.J.; Travers, P.; McDicken, I.W.: Demonstration of placental and placental-like alkaline phosphatase in non-malignant tumour tissue extracts using monoclonal antibodies in an enzyme immunoassay. Clinica chim. Acta *137:* 341–348 (1984).

9 Epenetos, A.A.; Snook, D.; Hooker, G.: Indium-111 labelled monoclonal antibody to placental alkaline phosphatase in the detection of neoplasms of testis, ovary and cervix. Lancet *ii:* 350–353 (1985).

10 Pawlikowska, T.R.B.; Epenetos, A.A.: Indium-111 labelled monoclonal antibody to placental-like alkaline phosphatase is of clinical value in the detection of neoplasms of the testis, ovary and cervix. Cancer Treat. Symp. (in press).
11 Fraker, P.J.; Speck, J.C.: Protein and cell membrane iodination with a sparingly soluble chloramide, 1,3,4,5-tetrachloro-5,6-diphenylglycouril. Biochem. biophys. Res. Commun. *80:* 849–857 (1978).
12 Pawlikowska, T.R.B.; Epenetos, A.A.: Clinical results with regional antibody guided irradiation. Cancer Treat. Symp. (in press).
13 Epenetos, A.A.: Antibody guided irradiation of malignant lesions. Three cases illustrating a new method of treatment. Lancet *i:* 1441–1443 (1984).

A.A. Epenetos, PhD, MRCP, Department of Clinical Oncology,
Hammersmith Hospital, London W12 0HS (UK)

Adv. clin. Enzymol., vol. 3, pp. 60–66 (Karger, Basel 1986)

Gamma-Glutamyltransferase Isoforms

From the Cell to the Plasma

G. Siest, A. Visvikis, A. Pawlak, B. Antoine, L. Wolff-Klein, N. Febvre, M.M. Galteau, M. Wellman-Bednawska

Centre du Médicament et Centre de Médecine préventive, Nancy, France

In clinical chemistry and biochemical pharmacology, gamma-glutamyltransferase (GGT) is a useful index for exclusion of hepatic disorders and effects of drugs (including alcohol). GGT is a membrane glycoprotein which is increased in cholestasis, carcinomas, and after administration of inducers such as phenobarbital. Its activity in plasma is highly correlated with plasma triacylglycerols and overweight. Conventional electrophoretic and chromatographic methods demonstrate the presence of different molecular forms in plasma and tissues [2]. These isoforms are not under separate gene control. Their use in different pathological or pharmacological situations requires a knowledge of the molecular and cellular bases of their synthesis and processing.

Tissue, Cellular and Subcellular Localization of GGT

In adult mammals, GGT is mainly found in kidney, pancreas, testis, mammary gland, intestine, brain and liver. In the fetus, liver GGT activity is high and after birth it decreases rapidly to adult values.

GGT is not homogeneously distributed within the liver. It is mainly located in the epithelium of bile ducts and at the biliary pole of hepatocytes. After administration of toxic compounds such as acetylaminofluorene, GGT is found in proliferating ductular cells, designated oval cells by Farber

[8]. We demonstrated after the group of Moss that GGT is also present in non-parenchymal cells, especially Kupffer cells [18]. It is an integral protein localized at the outside surface of the cells.

Because of its increase after inducing treatments, GGT, like cytochrome P-450, was thought to be located in microsomal fractions by many authors. However, using subcellular fractionation, we showed that, in rat and in rabbit, liver GGT is localized preferentially in plasma membranes [18]. The enzyme is also found in nuclear fractions due to contamination, not only by plasma membranes but also by biliary ductules, which occurs when the usual ultracentrifugation methods are used for liver fractionation [15].

In liver, about 90% of GGT activity is membrane bound. Imipramine and some other drugs produce an increase of soluble GGT [20]. After induction, the solubilization of GGT by detergents or biliary acids is different [18]. The solubilization in vivo by bile acids, drugs or proteases leads to several isoforms in plasma.

Structure

GGT of rat kidney is now well known. It is an amphipathic heterodimer composed of two glycosylated subunits. The heavier, H (M_r: $51 \cdot 10^3$) is anchored in the membrane via the N-terminal hydrophobic sequence (52 amino acids). The light subunit, L (M_r: $25 \cdot 10^3$) is linked to H by a non-covalent binding and supports the active site of the enzyme. The binding to the membrane through the N-terminal site is not very common: nevertheless sucrase-isomaltase, NADPH cytochrome P-450 reductase, cytochrome P-450, aminopeptidase are bound in that way.

The amino-terminal portion of the heavy subunit comprises 22 hydrophobic amino acids and resembles the structure of the signal peptide of the secretory proteins [11]. Actually, the N-terminal amino acid is methionine and this portion contains positively charged amino acids such as lysine and arginine, followed by hydrophobic amino acids.

Guellaën et al. [7], in Paris, are confirming this structure by cloning the rat kidney enzyme, by screening a cDNA library prepared from rat kidney mRNA enriched in the precursor message. The complete amino acid sequence of the protein was deduced from the nucleotide sequence of two clones. Guellaën et al. [7] found the same membrane hydrophobic sequence published by Matsuda et al. [11] and this result definitively demonstrated

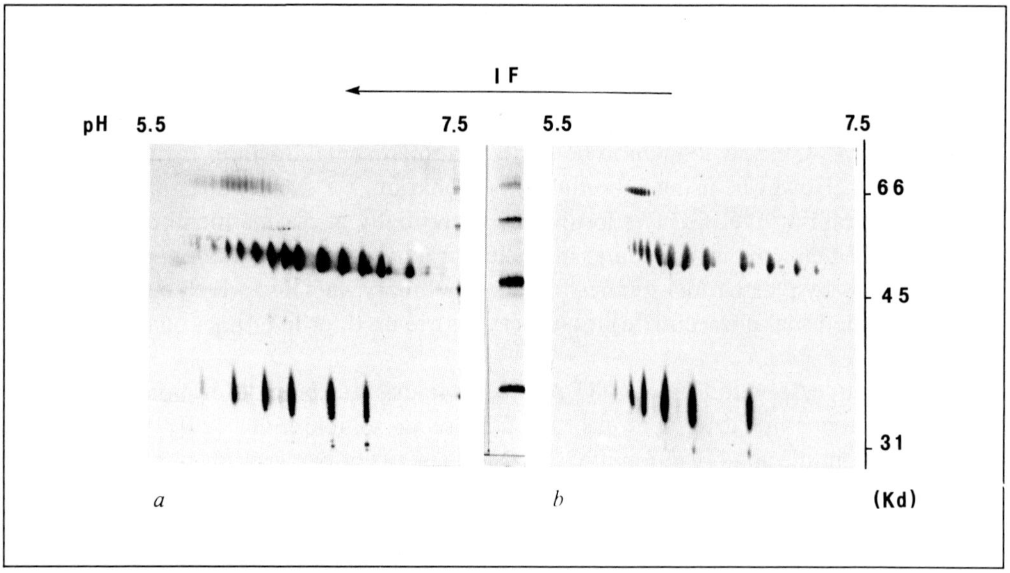

Fig. 1. Two-dimensional electrophoretic profile of papain-purified kidney GGT. *a* Silver staining. *b* Immunoblotting procedure. IF = Isoelectrofocalisation; Kd = kilodaltons.

that GGT is anchored in the plasma membrane via an uncleaved signal sequence.

The glycoprotein nature of GGT is now well established. The differences in the structure of the sugar chain are responsible for many different isoforms. Tate and Meister suggested that the 12 enzymatically active enzymes separated from purified rat kidney GGT by isoelectrofocusing are differently sialylated. Indeed, the sialic acid content of these forms varies from 14 to 61 nmol/mg protein. We recently obtained a nice demonstration of the heterogeneity of the papain-purified renal enzyme [16] by two-dimensional gel electrophoresis. We found 12 different heavy subunits but also 6 light subunits and several isoforms for the precursor. Each molecular form was recognized by polyclonal antibody after immunoblotting. These results confirmed that the precursor is glycosylated (fig. 1).

The sugar chains are all asparagine linked. In purified adult human GGT liver, we found that sialic acid content varied from 19 to 180 nmol/mg protein and represented, after DEAE cellulose separation, 8 multiple forms [18].

GGT purified from a hepatoma was found to possess a higher sialic acid content. Using concanavalin A, Köttgen proposed to separate the sialic acid-rich fetal forms from the sialic acid-poor adult forms. However, the sugar chain is much more complicated than is shown by the simple estimate of sialic acid amount and this approach is no longer used.

Liver GGT is less well understood than kidney enzyme. This may be due to the difficulties arising from the low GGT activity present in the liver and some differences found in polypeptides molecular weights. We are trying to obtain more information by following the same strategy as Guellaën et al. [7] but using the phage λ GT11 expression vector.

Biogenesis

Recently, four groups in New York, Philadelphia, Tokushima and Créteil contributed to important progress in the biogenesis and processing of the rat renal enzyme, both in in vitro and in in vivo systems [3, 5, 6, 11, 13, 14]. These authors showed that the two subunits of the kidney enzyme are first synthesized as a single and partially glycosylated precursor. They did not agree, however, on the exact point where the precursor is cleaved into two subunits. The precursor might already be cleaved in the endoplasmic reticulum (ER) or the cleavage could occur only after the transfer of the enzyme to the Golgi apparatus. In all cases, proteolysis and glycosylation would proceed along with the migration of the protein. Thus, as generally assumed, this membrane protein destined for the cell surface uses a pathway similar to that of secreted proteins. The studies mentioned above were done in the kidney except one which was performed in vitro in HTC cells, a rat hepatoma-derived cell line [4].

The isolation of the precursor is under way. What is important to realize is the very low enzymatic activity of the propeptide: specific activity is only 2% compared to that of the dimeric protein [19].

In kidney, the cleavage of the precursor begins in the ER. Matsuda et al. [12] are seeking two endoproteases involved in the processing of the kidney enzyme. However, in all studies a small part of the propeptide is found in the final location, i.e. the brush border membrane [9]. These authors tried to characterize high molecular weight forms consistently observed with the highly purified enzyme. They found that they are multimers (homo and heterodimers) formed by aggregation after denaturation resulting in new conformation and, of course, in supplementary isoforms.

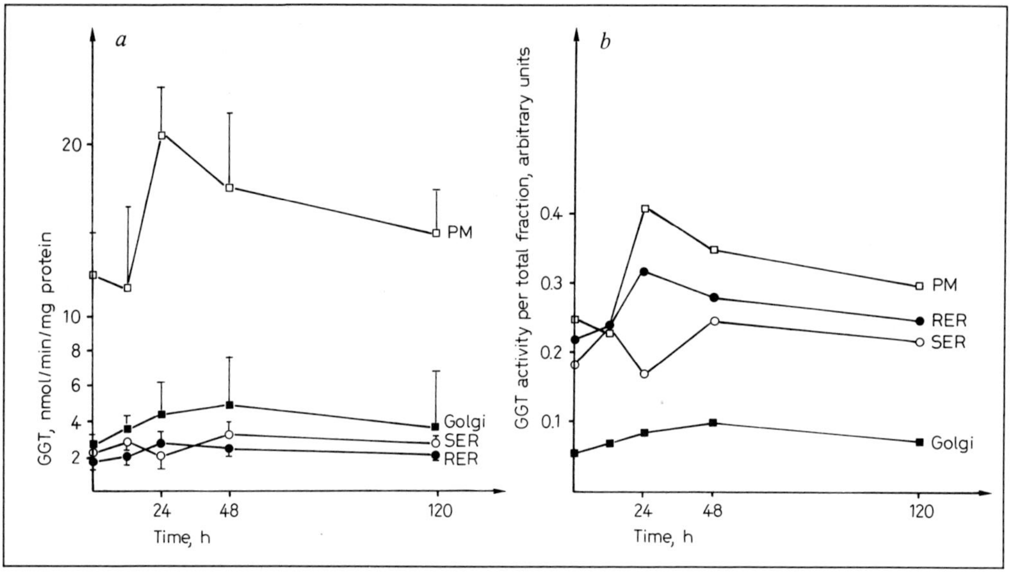

Fig. 2. a Effect of a single phenobarbital injection (75 mg/kg of body weight) on the evolution of GGT as a function of time in different subcellular membranes. *b* The distribution of the enzyme when the values are expressed on a total protein basis (i.e. the activity of each fraction if isolated from 100 mg protein of homogenate).

Induction

Phenobarbital, alcohol and aflatoxin are classical inducers of liver GGT [18]. Many structurally different drugs are capable of increasing GGT but very often the dose and time course are not strictly adapted to each compound and to each animal. We made a comparison of GGT induction and traffic in the different liver subcellular membranes. The time course of enzyme induction with phenobarbital (one dose: 75 mg/kg) showed a maximum increase 24 h after treatment in both the rough endoplasmic reticulum and the plasma membranes. This pattern differed from the distribution of drug metabolizing enzymes in subcellular membranes [1] (fig. 2).

After administration of a single critical dose of aflatoxin B_1 (one dose: 3 mg/kg) to male rats (Sprague Dawley) GGT activity already increased after 2 days in liver endoplasmic reticulum, Golgi apparatus and plasma membranes. Maximum increase was found 9 days after the intraperitoneal injection and reached 400–650% in the various endomembranes. Increased activities were also found in nuclei-enriched fractions (Wolff-Klein, unpublished data).

The comparison of two-dimensional electrophoretic profiles of each subcellular membrane of control and treated rats showed many sensitive variations induced by drugs (appearance or increase of numerous spots) [1, 16].

In plasma membranes, the heaviest subunit of hepatic GGT (M_r = $51 \cdot 10^3$) was identified as a succession of spots which differ by their PI. These spots were fewer and of lower apparent molecular weight in the ER. This suggests that some GGT precursor is already split at the ER level but is still not completely mature (glycosylated). A 5-day phenobarbital treatment induced the appearance of new spots in the plasma membranes [16].

The liver seems to be the only target organ of pharmacological induction. Very recently, Rambabu et al. [17] demonstrated that after alcohol administration and acivicin inhibition, GGT increased only in rat liver or serum but not in pancreas, brain, intestine, or kidney. So it seems that after induction, plasma GGT originated from liver. To study the biochemical mechanism at the cellular level, hepatoma cells are also very useful. Rat, mouse and human cell lines can be used to follow the induction phenomena.

In conclusion, all the molecular events in the cells can lead to the appearance of many isoforms of GGT: de-repression, proteolysis, intracellular traffic, recycling and induction are some of them. Pathological events (cholestase, carcinoma, diabetes) may also increase the heterogeneity of GGT. In addition, the hydrophobicity of these proteins underlies many possible associations with other proteins present in plasma [2].

References

1 Antoine, B.; Rahimi-Pour, A.; Siest, G.; Magdalou, J.; Galteau, M.M.: Differential time-course of induction of rat liver gamma-glutamyltransferase and drug-metabolizing enzymes in the endoplasmic reticulum, Golgi and plasma membrane after a single phenobarbital injection (Evaluation of proteins variations by two-dimensional electrophoresis). Cell. Biochem. Function (submitted).

2 Artur, Y.; Wellman-Bednawska, M.; Siest, G.: Formes moléculaires multiples de la gamma-glutamyltransférase plasmatique: caractérisation et signification clinique. Annls Biol. clin. *42:* 115–122 (1984).

3 Barouki, R.; Finidori, J.; Chobert, M.N.; Aggerbeck, M.; Laperche, Y.; Hanoune, J.: Biosynthesis and processing of gamma-glutamyltranspeptidase in hepatoma tissue culture cells. J. biol. Chem. *259:* 7970–7974 (1984).

4 Barouki, R.; Chobert, M.N.; Finidori, J.; Aggerbeck, M.; Nalpas, B.; Hanoune, J.: Ethanol effect in a rat hepatoma cell line: induction of gamma-glutamyltransferase. Hepatology *3:* 323–329 (1983).

5 Capraro, M.A.; Hughey, R.P.: Processing of the propeptide form of rat renal gamma-glutamyltranspeptidase. FEBS Lett. *157:* 139–143 (1983).
6 Finidori, J.; Laperche, Y.; Haguenauer-Tsapis, R.; Barouki, R.; Guellaën, G.; Hanoune, J.: In vitro biosynthesis and membrane insertion of gamma-glutamyltranspeptidase. J. biol. Chem. *259:* 4687–4690 (1984).
7 Guellaën, G.; Bulle, F.; Tabet-Aoul, T.; Aggerbeck, M.; Chobert, M.N.; Hanoune, J.; Laperche, Y.: Cloning and sequence of rat kidney gamma-glutamyltranspeptidase cDNAs. GGT Workshop, Pont d'Oye 1985.
8 Hixson, D.C.; Allison, J.P.: Monoclonal antibodies recognizing oval cells induced in the liver of rats by N-2-fluorenylacetamide or ethionine in a choline-deficient diet. Cancer Res. *45:* 3750–3760 (1985).
9 Hughey, R.P.: Heterogeneity in the post-translational processing of gamma-glutamyltransferases. GGT Workshop, Pont d'Oye 1985.
10 Inoue, M.; Hirata, E.; Morino, Y.: Regulation of hepatic turnover of gamma-glutamyltransferase and other ectoenzymes on apical membrane surface by bile acids. GGT Workshop, Pont d'Oye 1985.
11 Matsuda, Y.; Tsuji, A.; Katunuma, N.: Membrane bound gamma-glutamyltranspeptidase: structure, biosynthesis and degradation; in Advances in enzyme regulation, vol. 21, pp. 103–119 (Pergamon Press, Oxford 1983).
12 Matsuda, Y.; Kuno, T.; Katunuma, N.: Proteolytic processing of the precursor of GTP on the brush border membrane. GGT Workshop, Pont d'Oye 1985.
13 Nash, B.; Tate, S.S.: Biosynthesis of rat renal gamma-glutamyltranspeptidase. J. biol. Chem. *257:* 585–588 (1982).
14 Nash, B.; Tate, S.S.: In vitro translation and processing of rat kidney gamma-glutamyltranspeptidase. J. biol. Chem. *259:* 678–685 (1984).
15 Pecker, F.; Prado-Fugueroa, M.; Thines-Sempoux, D.; Amar-Costesec, A.: Membrane topology of gamma-glutamyltransferase and other enzymes related to the plasma membrane of rat liver cells. GGT Workshop, Pont d'Oye 1985.
16 Rahimi-Pour, A.; Wellman-Bednawska, M.; Galteau, M.M.; Artur, Y.; Siest, G.: Identification of gamma-glutamyltransferase in rat liver plasma membranes after two-dimensional electrophoresis. Electrophoresis *7:* 83–88 (1986).
17 Rambabu, K.; Matsuda, Y.; Katunuma, N.: Studies on turnover rates of rat gamma-glutamyltranspeptidase after chronic ethanol administration in vivo. Biochem. Med. (in press).
18 Siest, G.; Ratanasavanh, D.; Galteau, M.M.; Wellman-Bednawska, M.: Drug regulation of gamma-glutamyltransferase, a plasma membrane glycoprotein; in Akoyunoglou, Evangelopoulos, Georatsos, Palaiologos, Trakatellis, Tsiganos, Cell function and differentiation, part A, pp. 409–421 (Liss, New York 1982).
19 Tate, S.S.: Single-chain precursor of renal γ-glutamyl transpeptidase. Isolation and characterization. Fed. Proc. *44:* 710 (1985).
20 Tazi, A.; Galteau, M.M.; Siest, G.: Effect of imipramine on hepatic gamma-glutamyltransferase in female rats. Interaction with contraceptives. Biochem. Pharmac. *29:* 2874–2876 (1980).

G. Siest, MD, Faculté des Sciences Pharmaceutiques et Biologiques,
BP 403, F–54001 Nancy Cedex (France)

Adv. clin. Enzymol., vol. 3, pp. 67–73 (Karger, Basel 1986)

Biochemical Characterization and Some Clinical Aspects of the Multiple Forms of Gamma-Glutamyltransferase

N.-E. Huseby

Institute of Medical Biology, University of Tromsø, Norway

Serum γ-glutamyltransferase (GGT) has been separated into several multiple forms using various techniques. Three major components can be obtained by gel filtration or agarose gel electrophoresis: one macroform ($M_r > 600{,}000$), one medium form (M_r 250,000–450,000) and one minor form with M_r 90,000 [1–3]. The two larger forms predominate in serum from patients with hepatobiliary diseases and have been shown to be complexes of the enzyme with lipoproteins and lipid material [1, 3–8]. The third and minor form is hydrophilic and may be derived from the lipid-bound enzyme by proteolysis [9].

The present contribution contains data from two investigations on the heterogeneity of the two larger forms and one on the presence of the minor form. Some clinical aspects of these data will be discussed.

Material and Methods

Serum samples were obtained from patients at the University clinic in Tromsø. The techniques used in enzyme activity measurements, separations of multiple forms and in immunoprecipitation, have been described in earlier publications [1, 9]. Both anti-IgA and anti-LpX were found to contain antibodies to GGT and had to be absorbed prior to use. Separations of serum on fast protein liquid chromatography (FPLC) were performed with the gel filtration column Superose 6, as described by the manufacturer (Pharmacia, Uppsala, Sweden).

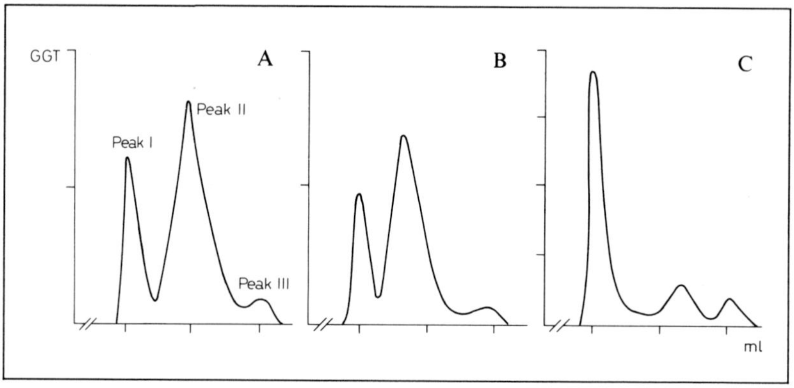

Fig. 1. Gel filtration of serum on Ultrogel Aca 34. Serum A (GGT activity 702 U/l) was from a patient with choledocholithiasis, serum B (GGT 540 U/l) and serum C (GGT 733 U/l) from 2 patients with biliary cancer.

Results and Discussion

Separation of GGT by Gel Filtration: Heterogeneity of Lipid-Bound GGT Complexes in Pathological Sera

Serum was sampled consecutively from 63 in-patients in the gastrosurgical department. Sera with increased GGT activity (> 55 U/l, 28 patients) were subjected to gel filtration. Figure 1 shows the elution profile of GGT activity for three of these sera. The amount of the minor form (peak III, fig. 1) was less than 10% of the total GGT activity for all sera examined. Peak II was the dominating fraction of GGT activity in most sera. When peak II eluates were incubated with antiserum to apolipoprotein A more than 80% of GGT activity was precipitated. As indicated in figure 1, there were variations in the elution profiles of this intermediary peak [1, 3]: however, all eluates apparently contained HDL-GGT complexes. The dominance of such complexes was also detected in another, larger group of patients with minor liver affections without icterus. This is in accordance with other reports [1, 3, 4, 10, 11].

The amount of activity eluting in the void volume (peak I) ranged from 10 to 80%. In sera with demonstrable LpX a large fraction of this peak is LpX-GGT complexes. Such complexes also contributed to peak I activity in several of the LpX-negative sera, ranging from 0 to 30%. Thus, in serum C (LpX-positive) 80% of the GGT activity eluted with peak I and 60% of

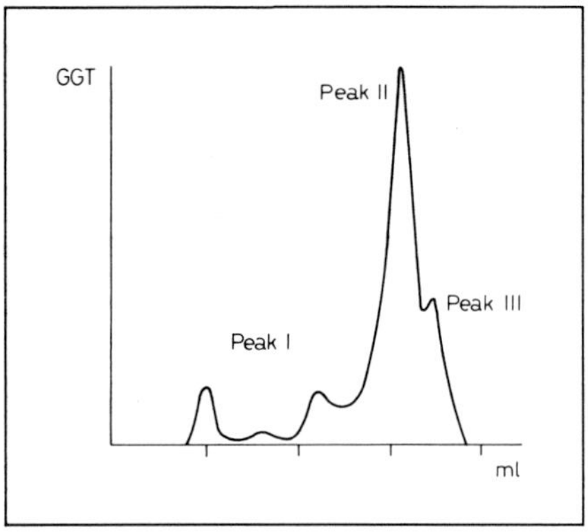

Fig. 2. Serum fractionated on FPLC. The elution profile of GGT is classified as group A; with a dominating peak II (HDL-GGT complexes). The serum was obtained from an alcoholic patient.

this peak was precipitated with anti-LpX. Both serum A and serum B were LpX-negative, and the relative size of peak I was the same (25–30%). LpX-GGT was estimated to contribute 25% of peak I activity in serum A whereas no LpX-GGT was demonstrated in serum B. When these data were compared to the diagnosis of the 3 patients (see legend to fig. 1), no clear correlation could be seen. Although a significant peak I activity appears frequently with icterus, the clinical value in differential diagnosis can be questioned [10, 11].

No precipitation could be detected using antiserum to apolipoprotein B or IgA. This is in contrast to other reports [4, 7, 8] but may be due to different antisera or precipitation techniques.

Separation of GGT-Lipid Complexes by FPLC: Improved Resolution

Serum from 24 patients with a variety of diseases were collected, the GGT activity ranged from 250 to 1,680 U/l. Based on the elution profiles after gel filtration on FPLC, these sera could be separated into three groups. However, the classification was ambiguous in several cases. Different classifications of two (or more) sera were also seen when following some patients for 1–2 weeks.

Figure 2 shows the profile of a serum from group A, which is characterized by one dominating peak. This peak is equivalent to peak II in the former separation system, and consists of GGT-HDL complexes. Thus, this appears to be a most common pattern. The minor peaks to the left of peak II were included in peak I in the former separation system, and will be designated peaks IA, IB, IC and ID. The patients in the present material showed mainly liver affections without icterus.

Figure 3 shows the elution profiles of two sera belonging to group B. In this group, peak II is still the dominating peak, but additional peaks appear more clearly; especially peaks IA and ID. The patients in this group apparently suffered from more severe liver affections than those in group A: hepatitis, biliary inflammation, icterus. The upper part of figure 3 shows the elution pattern of serum from a patient with drug-induced hepatitis. This pattern was almost constant for the 2 weeks of hospitalization; peak ID was slightly increased in the earlier period. The other serum in figure 3 was sampled from a cancer patient with liver metastases and evolving icterus. Peak IA amounted to 35% of the total activity, and a third of the activity in this peak was precipitated with antibodies to LpX. The other large M_r forms were not precipitated with antibodies to LpX, apoB, apoA or apoC. When analysing the densities of these large complexes in ultracentrifugation, a further heterogeneity was indicated. A varying amount (30–60%) of GGT in these peaks floated at density 1.06, the rest at 1.22. The nature of these large complexes could not be established but might be some phospholipid vesicles.

Also indicated in figure 3 is the elution profile of alkaline phosphatases (ALP). This enzyme is also known to be part of high molecular mass complexes [4, 5, 7]. Both ALP and GGT have been identified in LpX complexes [4, 7] and these as well as other plasma membrane enzymes have been proposed to be part of plasma membrane fragments ('koinozymes') in the circulation [5]. However, no activity of ALP was precipitated when peak I eluates were incubated with antibodies raised against purified liver GGT, which removed more than 90% of the GGT activity in peak I. Furthermore, no correlation between the relative amounts of high M_r forms of GGT and ALP or between the amount of GGT and ALP precipitated by anti-LpX was found. Thus, these data do not support the idea of 'koinozymes'.

Two sera classified in group C are shown in figure 4. This group (of 6 patients) is characterized by a very low peak II (HDL-GGT complexes), and large amounts of peak IA and/or peak ID. The patients in this group were suffering from severe liver diseases: 5 had liver metastases. The sera with large peak IA in this group also contained significant amounts of LpX-GGT. As found for sera in group B, no indications for GGT complexes with

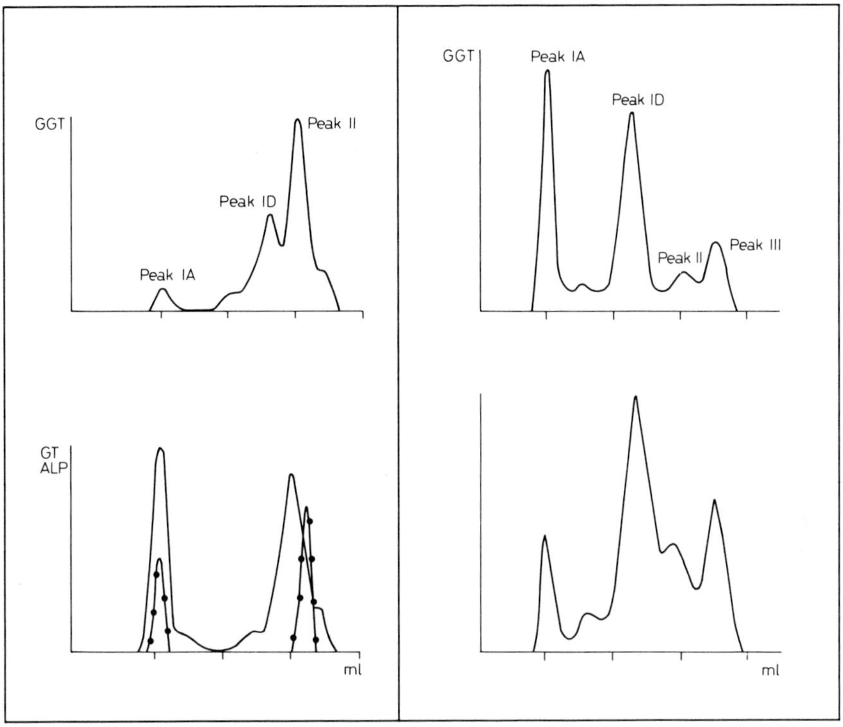

3 4

Fig. 3. Serum fractionated on FPLC. The elution profile of GGT is classified as group B, with a major peak II and significant peaks ID (upper graph) and/or peak IA (lower graph). In the lower graph ALP activity is also indicated (●).

Fig. 4. Serum fractionated on FPLC. The elution profile of GGT is classified as group B, with peak IA or peak ID being the major peaks. The 2 patients had liver metastases.

apolipoproteins were found for the remaining activity in peak IA or in peak ID. Furthermore, heterogeneity with respect to density was demonstrated for these macropeaks.

Increase of Peak III (Hydrophilic Form of GGT) in Paracetamol Intoxication

In the lower part of figure 4 an increased amount of the minor form of GGT (about 25%) can be detected. In figure 5, two graphs are shown of sera sampled from a patient who died with liver necrosis after paracetamol intoxication. The samples were collected 8 and 2 days prior to death. GGT

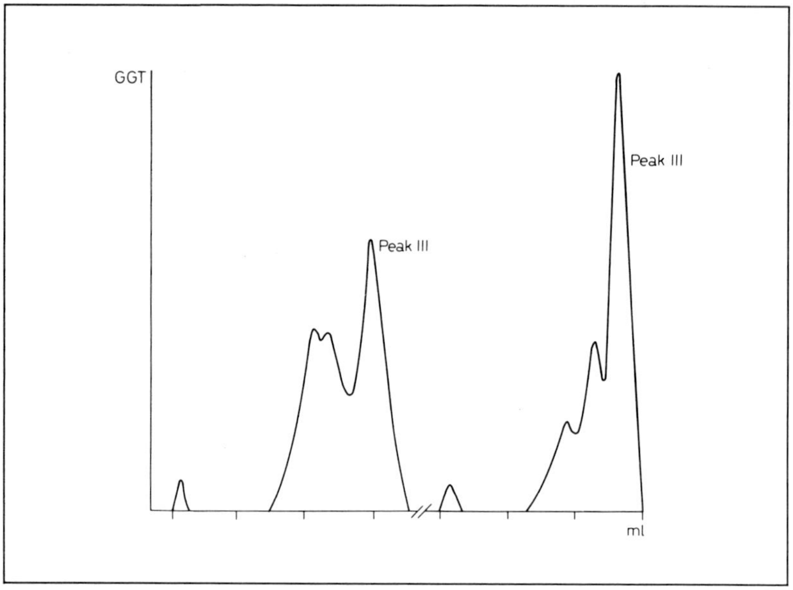

Fig. 5. Serum fractionated on FPLC. The graph shows two sera from a paracetamol-intoxicated patient. The dominating peak being peak III, the hydrophilic form of GGT.

activity was 150–130 U/l, and the amount of peak III is clearly increasing. A similar pattern was detected for another patient with fatal paracetamol intoxication [9]. As this minor or hydrophilic form of GGT can be produced by proteolytic activity in liver, the presence of increased peak III can be an indicator of liver necrosis. However, the GGT activities were low in these 2 patients when compared to other serum enzymes (aspartate aminotransferase and lactate dehydrogenase, both being 10,000–20,000 U/l). Besides, the amount of the hydrophilic GGT in most sera is low, also from patients with suspected liver necrosis.

An explanation might be that the hydrophilic enzyme is distributed in bile and plasma and also, due to its lower M_r, in lymph and interstitial fluid. In pig lymph, no high M_r form of GGT could be detected, but significant amounts of the hydrophilic enzyme were seen. Thus, when released from liver, the membrane-bound enzyme will be confined to plasma in the form of high M_r complexes, whereas the hydrophilic form is distributed to larger volumes and therefore may not reach the same high levels in plasma as the lipid-bound form.

Conclusions

The separation of multiple forms of GGT can be highly improved by FPLC. A major fraction of GGT in serum is bound to HDL. In more severe liver diseases (with icterus), the presence of several high M_r complexes of GGT can be detected. One such complex is that of LpX and GGT. The presence of the hydrophilic form in increased amounts may indicate liver necrosis. However, the clinical value of these separations is not established.

References

1 Huseby, N.E.: Multiple forms of serum γ-glutamyltransferase. Association of the enzyme with lipoproteins. Clinica chim. Acta *124:* 103–112 (1982).

2 Echetebu, Z.O.; Moss, D.W.: Multiple forms of human γ-glutamyltransferase. Preparation and characterization of different molecular weight fractions. Enzyme *27:* 1–8 (1982).

3 Wenham, P.R.; Horn, D.B.; Smith, A.F.: Physical properties of γ-glutamyltransferase in human serum. Clinica chim. Acta *141:* 205–218 (1984).

4 Arthur, Y.; Wellman-Bednawska, M.; Jacquier, A.; Siest, G.: Complexes of serum gamma-glutamyltransferase with apolipoproteins and immunoglobulin A. Clin. Chem. *30:* 631–633 (1984).

5 Broe, M.E. de; Borgers, M.; Wieme, R.J.: The separation and characterization of liver plasma membrane fragments circulating in the blood of patients with cholestasis. Clinica chim. Acta *59:* 369–372 (1975).

6 Wieland, H.; Bernard, R.; Baggio, G.; Müller, P.; Seidel, D.: Demonstration of a lipoprotein-GGT complex in the plasma of cholestatic patients. Eur. J. clin. Invest. *7:* 243 (1977).

7 Crofton, P.M.; Smith, A.F.: High molecular-mass alkaline phosphatase in serum and bile: nature and relationship with lipoprotein-X. Clin. Chem. *27:* 867–874 (1981).

8 Sudo, K.; Kano, T.: Membrane associated enzyme complexes containing lambda type of immunoglobulin A. Clinica chim. Acta *107:* 49–58 (1980).

9 Huseby, N.E.: Hydrophilic form of γ-glutamyltransferase: proteolytic formation in liver homogenates and its estimation in serum. Clinica chim. Acta *124:* 113–121 (1982).

10 Arthur, Y.; Wellman-Bednawska, M.; Jacquier, A.; Siest, G.: Associations between serum gamma-glutamyltransferase and apolipoproteins: relationships with hepatobiliary diseases. Clin. Chem. *30:* 1318–1321 (1984).

11 Wenham, P.R.; Horn, D.B.; Smith, A.F.: Multiple forms of γ-glutamyltransferase: A clinical study. Clin. Chem. *31:* 569–573 (1985).

N.-E. Huseby, MD, University of Tromsø, Institute of Medical Biology,
N–9001 Tromsø (Norway)

Adv. clin. Enzymol., vol. 3, pp. 74–82 (Karger, Basel 1986)

Multiple Forms of Gamma-Glutamyltransferase in Bile and Serum

P.R. Wenham

Department of Clinical Chemistry, Western General Hospital, Crewe Road, Edinburgh, UK

It is generally accepted that the multiple forms of gamma-glutamyltransferase (GGT; EC 2.3.2.2) in different tissues are not true isoenzymes, but represent post-translational modifications, such as the addition of sialic acid residues, of a single gene product [1, 2]. To date, a large number of GGT forms have been described in bile and serum by various workers using a wide range of physicochemical techniques [3–6]. Few reports, however, have attempted to interrelate the multiple forms obtained by using the various techniques, resulting in a literature that is somewhat complicated and confusing. The aim of this report is to attempt to paint a unifying picture of the multiple forms of GGT in bile and serum, using the results of other workers together with my own findings.

The nomenclature I shall use is based upon the fractions obtained by gel chromatography. A fairly clear pattern of results emerges whether serum or bile is used, and to simplify the description, the peaks are numbered with Arabic numerals, in order of their elution from the gel column. To maintain consistency, the corresponding electrophoretic fractions are numbered with the appropriate Roman numeral. The fractions are:

Peak 1: void volume peak ($M_r > 1{,}000{,}000$) present in both serum and bile which may be very heterogeneous.

Peak 2: major peak of intermediate M_r (250,000–500,000) which in serum can take two forms (see later). This peak is not present in native bile.

Peak 3: peak of low M_r (about 160,000) found in bile when bile salts are present in the elution buffer.

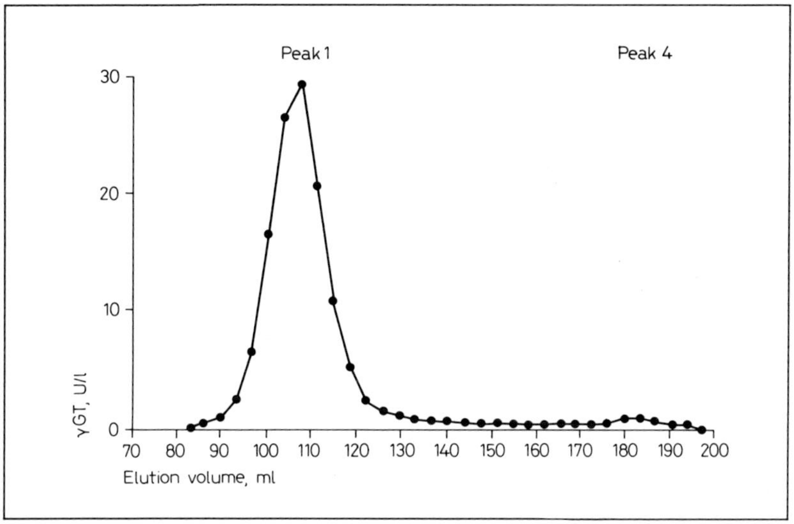

Fig. 1. Elution profile of GGT in human bile following gel chromatography on Sephadex G200.

Peak 4: peak of low M_r (about 120,000) present as a minor fraction in both serum and bile. Its size is increased greatly after treatment of both bile and serum with papain or trypsin.

The nature of the enzyme present in the bile and in serum appears to be different and so each will be considered separately.

Nature of GGT in Bile

Following gel chromatography, over 85% of GGT in hepatic bile elutes as peak 1, with the remainder as peak 4 (fig. 1) [7, 8]. However, in the presence of a concentration of deoxycholate that is 5 mmol/l or greater, or of physiological concentrations, namely 20 mmol/l or greater of glycocholate or 5 mmol/l or greater of glycochenodeoxycholate, the peak 1 form is converted to the peak 3 form [1, 7, 8]. This molecule is hydrophobic in nature and appears to reaggregate if bile salts are removed either by dilution or dialysis. It is believed to consist of a GGT-detergent complex, conferring on the molecule a greater negative charge and consequently, faster electrophoretic mobility than the peak 4 form even on seiving media. In contrast, the latter is hydrophilic in nature, forming neither complexes nor aggregates when added to either bile or serum [7]. After high speed centrifugation of bile, between 25 and 75% of GGT activity remains in the supernatant. This suggests that a significant proportion of GGT is present as a soluble form. Moreover, this proportion is higher in bile samples with high bile salt concentrations and is also increased by prior addition of bile salts to the bile [8]. In contrast, the enzyme that sediments probably represents GGT in a particulate or reaggregated form.

Origin of GGT in Bile

Although most of the GGT in native bile appears to be present as a high M_r form, the technique involves considerable dilution of the samples in aqueous solution. This in effect reduces the bile salt concentrations in the solution containing the GGT. However, if the analytical system is set up so that the GGT is maintained in an aqueous solution containing physiological concentrations of bile salts, the enzyme is present largely as a low M_r form. Furthermore, the high and low M_r forms are interconvertible, depending on the bile salt concentration of the solution in which they are present. After incubating isolated hepatocytes in solutions of bile salts, membrane bound enzymes are released into the incubation medium [9]. At low bile salt concentration the enzymes appear mainly in a particulate form, whereas, at a higher concentration of bile salts the proportion released as a soluble form increases. In the light of these findings, it is possible to envisage the release of GGT into the bile, from the cells lining the biliary tract, by the bile salts present in the bile. At a low bile salt concentration the enzyme is present as particulate material possibly in the form of small vesicles released by the hepatocytes or pinched off the tips of the microvilli of the canalicular membrane [10]. However, when the concentration of bile salts increases, it is leached from the particles and passes into solution.

Nature of GGT in Serum

Varying amounts of GGT elute as the peak 1, peak 2 and peak 4 forms following gel chromatography of serum. The peak 2 form exhibits an apparent dichotomy [11]. In patients with extrahepatic obstruction it tends to

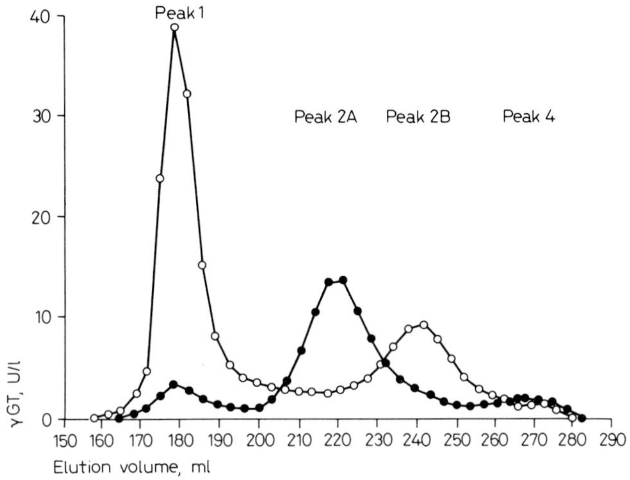

Fig. 2. Elution profile of GGT following gel chromatography on Sephacryl S300 of serum from 2 patients with liver disease. ○ = Extrahepatic biliary obstruction; ● = alcoholic cirrhosis.

elute later (peak 2B) than in either normal individuals or in patients with other forms of liver disease (peak 2A) (fig. 2). The peak 1 and both of the peak 2 forms appear to be hydrophobic in nature. They are converted by bile salts to the peak 3 form and by papain digestion to the peak 4 form.

Following treatment of serum with polyanions, between 40 and 75% of peak 1 activity co-precipitates with the LDL and VLDL fraction [11]. This is consistent with previous reports that high M_r GGT may, in part, consist of complexes of the enzyme with lipoproteins or with lipoprotein-X [12–15]. In contrast, between 49 and 65% of peak 2A activity co-precipitates with the HDL fraction after the further addition of polyanions to serum [11], supporting the previous suggestions that the intermediate M_r fraction consists of a complex of GGT and HDL [13]. Further evidence for this hypothesis is that varying amounts of serum GGT activity are precipitated by antiserum to apolipoprotein A but not by antiserum to apolipoprotein B [11, 13]. Electrophoretic studies confirm that the antiserum to apolipoprotein A binds to the GGT fractions of intermediate M_r (bands IIA and band IIB) (fig. 3).

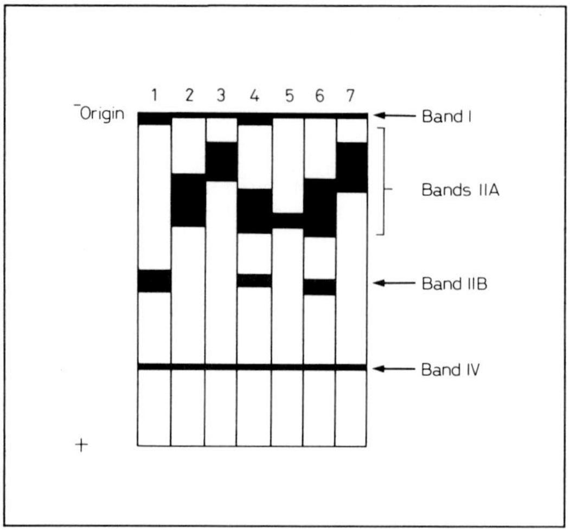

Fig. 3. Schematic representation of the various zones of GGT activity obtained after polyacrylamide slab gel electrophoresis of sera from patients with liver disease. 1 = Extrahepatic biliary obstruction; 2 = alcoholic cirrhosis; 3 = chronic active hepatitis; 4 = primary biliary cirrhosis; 5 = haemochromatosis; 6 = alcoholic cirrhosis; 7 = liver metastases.

Clinical Aspects of the Multiple Forms in Serum

In an attempt to improve the diagnostic efficiency of GGT measurements, and in particular, to aid the differentiation between 'surgical' and 'medical' causes of jaundice, two of the multiple forms of GGT have been studied in detail, namely peak 1 and peak 2B [16]. The former has been measured, as a percentage of total activity, in 100 patients with liver disease, using gel chromatography (peak 1(%)). The latter has been estimated, as the percentage of the total intermediate M_r GGT, in 64 patients, by polyacrylamide gradient gel electrophoresis and densitometric scanning (% band IIB).

Peak 1(%): (1) is greatest in patients with extrahepatic obstruction ($p < 0.000001$); (2) is greater in patients with liver metastases than in other groups, after the exclusion of patients with extrahepatic obstruction ($p < 0.0001$), and (3) is greater in patients with extrahepatic obstruction than in jaundiced patients in other disease categories ($p < 0.00001$).

Band IIB reached a value of 100% in 11 out of 16 patients (69%) with obstructive jaundice, an observation that could also be noted by simple

Table I. Diagnostic value of peak 1 (%) and % band IIB in the detection of extrahepatic biliary obstruction

	Peak 1	Band IIB
Sensitivity[1]	11/16 (69)	14/16 (88)
Specificity[2]	44/48 (92)	46/48 (96)
Predictive value of positive test[3]	11/15 (73)	14/16 (88)
Predictive value of negative test[4]	44/49 (90)	46/48 (96)

[1] Proportion of patients with extrahepatic obstruction giving a positive test.
[2] Proportion of patients without extrahepatic obstruction giving a negative test.
[3] Proportion of patients with a positive test who had extrahepatic obstruction.
[4] Proportion of patients with a negative test who did not have extrahepatic obstruction.

inspection of the gels. In none of the patients of any other group was all of the intermediate M_r GGT present as band IIB. A comparison of the diagnostic value of peak 1(%) and % band IIB in the detection of extrahepatic obstructive jaundice is shown in table I. A positive test is defined as a value greater than the mean + 2 SD for each variable in the patients without obstructive jaundice. Whereas both tests have similar specificities, the measurement of band IIB is much more sensitive in the detection of obstructive jaundice than is the measurement of peak 1(%).

Origin of GGT in Serum

The exact relationship between biliary and serum GGT remains obscure, as does the mechanism whereby the serum activity rises in patients with liver disease. In cholestatic lesions, due to biliary regurgitation, an increased amount of biliary enzyme (peak 3) might find its way into the circulation. There, surrounded by a lower bile salt concentration, it would be able to form hydrophobic bonds with lipids or lipoproteins and generate the complexes described above. However, a significant proportion of biliary GGT may be particulate in nature [8]. It is therefore possible that the appearance of membrane fragments (to which GGT and other hepatocyte plasma membrane enzymes are attached) in the circulation of patients with cholestasis [17] is due to the regurgitation of particulate material from bile into the circulation. In patients with alcoholic liver disease or on long-term anticonvulsant therapy, recent evidence has suggested the possible occurrence of two processes. Firstly, increased hepatic synthesis of the enzyme and secondly an alteration in the chemical composition of the membranes rendering their constituents more susceptible to bile sale-mediated release [18, 19]. In conclusion, therefore, it seems likely that bile salts play a part, in one way or another, in the release of GGT into the circulation in liver disease. However, the exact processes which facilitate their interaction with the hepatocyte remain to be determined.

GGT as a Liver Function Test

The differential diagnosis of jaundice is a common diagnostic problem in routine clinical practice. Ideally, a test is required that is readily available, rapid, safe and accurate. At present, however, there is no single technique that satisfies all of these criteria. More complex analysis of serum GGT patterns has increased its discriminatory capacity. In particular, either the measurement of peak 1(%) or % band IIB may help to distinguish extrahepatic obstruction: (a) from intrahepatic obstruction, and (b) from almost every other form of liver disease. However, measurement of the former is time-consuming and tedious and is not suited for routine use. In contrast, serum GGT electrophoresis is a fairly straightforward technique that is suited to most laboratories. However, the present study had several limitations. Firstly, it was not possible to follow the various diseases over their natural clinical course. Secondly, the number of patients studied was small, and, thirdly, as only a single specimen was obtained from each patient, many patients were studied at different points in time during the course of their disease. For example, patients with extrahepatic obstruction were studied soon after the onset of jaundice whereas some of the patients with jaundice due to other causes were not studied until they had been jaundiced for some time. It is possible that these factors may have had a bearing upon the GGT patterns obtained. A 'longitudinal' study of patients is required, whereby each patient is looked at: (a) as soon after the onset of the disease as possible, and (b) regularly throughout the course of the disease. Such a study would demonstrate the value of GGT electrophoresis and whether or not it could assist in rendering the more expensive and invasive clinical tests unnecessary.

Acknowledgements

I thank Dr. D.B. Horn and Dr. A.F. Smith for their advice in the preparation of this manuscript.

References

1 Huseby, N.E.: Multiple forms of γ-glutamyltransferase in normal human liver, bile and serum. Biochim. biophys. Acta *522:* 354–362 (1978).
2 Echetebu, Z.O.; Moss, D.W.: Multiple forms of human γ-glutamyltransferase. Preparation and characterization of different molecular weight fractions. Enzyme *27:* 1–8 (1982).
3 Hetland, O.; Anderson, T.R.; Gerner, T.: The heterogeneity of the serum activity of γ-glutamyltranspeptidase in hepatobiliary diseases as studied by agarose gel electrophoresis. Clinica chim. Acta *62:* 425–431 (1975).
4 Burlina, A.: Improved method for fractionating γ-glutamyltransferase by electrophoresis on cellulose acetate. Clin. Chem. *24:* 502–504 (1978).
5 Kokot, F.; Kuska, J.: Heterogeneity of serum γ-glutamyltranspeptidase in different internal diseases studied by starch-gel electrophoresis and Sephadex filtration. Enzymol. biol. clin. *9:* 59–67 (1968).
6 Wenham, P.R.; Price, C.P.; Sammons, H.G.: γ-Glutamyltransferase isoenzymes in human bile. J. clin. Path. *31:* 666–670 (1978).
7 Wenham, P.R.; Horn, D.B.; Smith, A.F.: γ-Glutamyltransferase in bile and sera from patients with extrahepatic biliary obstruction. Clinica chim. Acta *112:* 113–122 (1981).
8 Wenham, P.R.; Horn, D.B.; Smith, A.F.: The nature of γ-glutamyltransferase and other hepatocyte plasma membrane enzymes in human bile. Clinica chim. Acta *124:* 303–313 (1982).
9 Billington, D.; Evans, C.E.; Godfrey, P.P.; Coleman, R.: Effect of bile salts on the plasma membranes of isolated rat hepatocytes. Biochem. J. *188:* 321–327 (1980).
10 Godfrey, P.P.; Warner, M.J.; Coleman, R.: Enzymes and proteins in bile. Variations in output in rat cannula bile during and after depletion of the bile salt pool. Biochem. J. *196:* 11–16 (1981).
11 Wenham, P.R.; Horn, D.B.; Smith, A.F.: Physical properties of γ-glutamyltransferase in human serum. Clinica chim. Acta *141:* 205–218 (1984).
12 Wenham, P.R.; Price, C.P.; Sammons, H.G.: Serum γ-glutamyltransferase isoenzymes in extra-hepatic biliary obstruction. J. clin. Path. *32:* 902–906 (1979).
13 Huseby, N.E.: Multiple forms of serum γ-glutamyltransferase. Association of the enzyme with lipoproteins. Clinica chim. Acta *124:* 103–112 (1982).
14 Crofton, P.M.; Smith, A.F.: High-molecular mass alkaline phosphatase in serum and bile. Nature and relationship with lipoprotein-X. Clin. Chem. *27:* 867–874 (1981).
15 Echetebu, Z.O.; Moss, D.W.: Multiple forms of human γ-glutamyltransferase. Modification and possible structures of different molecular weight fractions. Enzyme *27:* 9–18 (1982).

16 Wenham, P.R.; Horn, D.B.; Smith, A.F.: Multiple forms of γ-glutamyltransferase. A clinical study. Clin. Chem. *31:* 569–573 (1985).
17 De Broe, M.E.; Borgers, M.; Wieme, R.J.: The separation and characterization of liver plasma membrane fragments circulating in the blood of patients with cholestasis. Clinica chim. Acta *59:* 369–372 (1975).
18 Ratanasavanh, D.; Tazi, A.; Galteau, M.M.; Siest, G.: Localisation of gamma-glutamyltransferase in subcellular fractions of rat and rabbit liver. Effect of phenobarbital. Biochem. Pharmac. *28:* 1363–1365 (1979).
19 Ratanasavanh, D.; Magdalou, J.; Antoine, B.; Galteau, M.M.; Siest, G.: Gamma-glutamyltransferase activity of liver plasma membranes in phenobarbital-treated rabbits. Parmacol. Res. Commun. *13:* 909–919 (1981).

P.R. Wenham, PhD, Department of Clinical Chemistry,
Western General Hospital, Crewe Road, Edinburgh EH4 2XU (UK)

Adv. clin. Enzymol., vol. 3, pp. 83–89 (Karger, Basel 1986)

Characterization of the Multiple Forms of Serum Gamma-Glutamyltransferase

The Use of Immunochemical Methods

Y. Artur, M. Wellman-Bednawska, G.J. Sanderink, R. Donati, N. Sabolovic, F. Schiele

Centre de Médecine Préventive et Centre du Médicament, Vandœuvre lès Nancy, France

According to Huseby [9], gamma-glutamyltransferase (GGT; EC 2.3.2.2.) is present in human organs as an amphiphilic membrane-bound enzyme. Detergents may solubilize the enzyme by binding to the hydrophobic, membrane-binding domain of the enzyme, forming a micellar-like complex, the 'heavy' or amphiphilic form of the enzyme. Treatment with proteases such as papain or trypsin removes the hydrophobic domain, thus forming a soluble, hydrophilic 'light' form of the enzyme. Both the heavy and the light forms are present in serum [11]. The heterogeneity of serum GGT mainly results from three types of phenomena occurring simultaneously [6]: variable degree of sialylation of the molecule, association with other proteins (lipoproteins, immunoglobulins, etc.), proteolytic modification of the enzyme in vivo or during storage of the samples.

The study of the immunological properties of the enzyme has been developed over about 10 years and many authors have used immunochemical methods in order to determine and characterize its multiple forms. We review here the various reports concerning the use of antibodies to GGT and to other proteins in the study of the multiple forms of serum GGT.

Use of Polyclonal Antibodies to GGT

Studies on the immunological reactivity of the various GGTs are based mainly upon double immunodiffusion tests (Ouchterlony's technique) and upon reactions of immunoprecipitation and immunoinhibition of the catalytic activity: indeed, the binding of the antibodies to the enzyme leads to an inhibition of its activity; nevertheless, all authors have observed that the immune complexes of GGT with antibodies retain a fraction of the activity of the native enzyme.

Double Immunodiffusion, Immunoprecipitation and Immunoinhibition of the Catalytic Activity of GGT

An immunological identity has been found by double immunodiffusion between the liver and kidney GGTs [8, 9, 17, 23, 25, 26, 29], and between the light (hydrophilic) and the heavy (amphiphilic) forms of the enzyme [5, 9]. In addition, GGTs from hepatoma [19, 29], fetal liver [5, 9], renal carcinoma [8], pancreas [9], serum [9, 25], bile [17], and urine [9, 23] are immunologically identical to the liver or kidney enzyme. The percentages of immunoinhibition of the catalytic activity of all these enzymes generally vary from 70 to 90% [5, 8, 12, 20, 29]; lower percentages were found by Szewczuk et al. [23] for the liver enzyme (57%) and by Tsuchida et al. [25] for liver and kidney GGT (40–50%) and for GGT fractions isolated from serum (20–50%). The percentages of immunoprecipitation obtained with the polyclonal antibodies reach 75–90% for the liver, kidney and serum enzymes [2, 5, 17, 20, 23, 25, 26]. In healthy subjects, as in patients with various liver diseases, the percentages of immunoprecipitation and immunoinhibition of the catalytic activity do not differ significantly [2, 5]. The results displayed above and many studies performed in animals (short review in Artur et al. [5]) suggest the conclusion that, within a mammalian species, GGTs from different organs and biological fluids are immunologically similar. This is in agreement with a post-translational origin of the heterogeneity of the enzyme, especially in serum. In conclusion, double immunodiffusion experiments, immunoinhibition and immunoprecipitation tests using polyclonal anti-GGT antibodies are without interest in the study of the multiple forms of the transferase.

Electroimmunodiffusion

Until now, little has been done on the immunochemical measurement of the GGT protein. Rocket electroimmunoassays were performed by Scherberich et al. [18] for the urine enzyme and by Huseby [9] for a partially purified liver GGT, but without quantitative application. A radioimmunoassay for human serum and pancreatic GGT [16] and enzyme-linked immunosorbent assays for the enzyme from some bovine body fluids and tissue preparations [22] and for GGT from rat liver and hepatoma [24] have

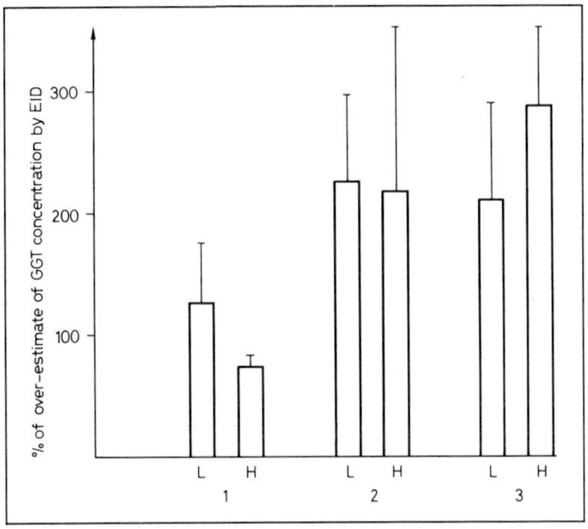

Fig. 1. Percentage of over-estimate of serum GGT concentration by electroimmunodiffusion (mean + standard deviation) (4). These percentages are calculated as follows: % = (P–Pn)/Pn. P and Pn are the GGT concentrations determined by electroimmunodiffusion (EID) before and after desialylation, respectively. Results are given for each of the two superimposed rockets (H and L) which are observed for all sera. 1: Patients with noncancerous or nonalcoholic liver diseases (viral hepatitis, congestive heart failure, lithiasis). 2: Patients with alcoholic steatosis or cirrhosis. 3: Patients with metastatic liver cancer.

recently been proposed. Using specific antibodies to the human kidney enzyme, our group has developed a rocket electroimmunoassay for human kidney and serum GGT [26]. Neuraminidase treatment of the samples has shown that the height of the rockets is highly influenced by the sialic acid content of the molecule. The peaks corresponding to the sialylated GGT are higher than those corresponding to the neuraminidase-treated enzyme. The differences in the peak heights before and after neuraminidase treatment of the sera have been studied in patients with various liver diseases [4]. They are significantly larger in patients with metastastic liver carcinoma or with alcoholic diseases than in patients with other noncancerous or nonalcoholic disorders (fig. 1). This is in agreement with the results obtained by others concerning the increased sialylation of the serum enzyme in patients with alcoholic diseases [15] or hepatic cancer [14]. Thus, electroimmunodiffusion seems to represent a promising technique for determining the degree of sialylation of GGT, which is increased in some physiological and pathological conditions (short review in Wellman-Bednawska et al. [26]).

Use of Antibodies Directed against Other Plasma Proteins

Since associations of GGT with other proteins represent one of the causes of its heterogeneity in serum, several authors have used antibodies in order to identify the proteins to which GGT is bound and to quantify these associations.

Use of Antibodies against (Apo)-Lipoproteins

There are several reports of binding of GGT to lipoproteins, in which the authors have used various techniques such as electrophoresis, ultracentrifugation and immunochemistry. Table I presents the results obtained by immunoelectrophoresis and immunoprecipitation in patients with liver diseases. Immunoprecipitation allows a fast and easy quantitation of the activity bound to lipoproteins. Thus, in sera of patients with liver diseases, apo-A-bound GGT activity represents 15–72% of total activity according to our group [3] and 8–92% according to Wenham et al. [27].

Our results suggest that the binding of the enzyme to this apo-A is not related to a clearly defined disease. Conversely, the apo-B-bound fraction of GGT is significantly increased in cholestasis, where it may represent up to 56% of the total serum activity [3]. Huseby [10] has separated the activity of

Table I. Association of GGT with lipoproteins in patients with liver diseases: immunochemical approach

Authors	Antiserum to	Reactivity	Method
Wieland et al. [28]	LpX	+	immunoprecipitation
Crofton and Smith [7]	LpX β-lipoprotein	+ +	immunoelectrophoresis and immunoprecipitation
Huseby [10]	α-lipoprotein apo B LpX	+ – +	immunoprecipitation
Wenham et al. [27]	apo A apo B	+ –	immunoelectrophoresis and immunoprecipitation
Artur et al. [2, 3]	apo A apo B	+ +	immunoprecipitation
Izumi [13]	α-lipoprotein β-lipoprotein	+ +	immunoelectrophoresis

GGT into two major fractions by gel filtration of pathological sera; antiserum precipitation has shown that alpha-lipoprotein-bound GGT represents 40–84% of the activity of one of the fractions, and lipoprotein X-bound GGT, 8–84% of the activity of the other. Complexes between GGT and lipids have generally been observed in pathological sera with high GGT activities, but Crofton and Smith [7] have found GGT-beta-lipoprotein complexes in sera from healthy subjects. However, the clinical significance of such associations is still under investigation.

Use of Antibodies to Immunoglobulins

Using a procedure of immunofixation with anti-IgA antibodies after electrophoresis, Sudo and Kanno [21] reported the presence, in some sera from patients with liver diseases (especially cholestasis), of high molecular-mass complexes containing alkaline phosphatase, arylamidase, GGT and IgA. In the same way, we have noted that an antiserum to IgA is able to precipitate a fraction of GGT activity in some sera of such diseased subjects. This fraction has never exceeded 22% of the total activity in the sera we have studied [2]; similarly, we have found in some patients a fraction of GGT activity bound to IgG [1]. We have never found Ig-GGT complexes in healthy subjects, but the pathophysiological significance of these complexes has not been elucidated. However, it is suggested that the occurrence of an immunological reaction (auto-antibodies) must be considered in the study of the Ig-GGT complexes.

Conclusion and Perspectives: The Use of Monoclonal Antibodies against GGT (Assay of GGT Epitopes)

Although heterogeneity of serum GGT seems to result from post-translational modifications of a single type of enzyme protein, the use of polyclonal antibodies against GGT has provided interesting information concerning one of the causes of its heterogeneity (the degree of sialylation). In addition, antibodies against lipoproteins and against immunoglobulins have provided better knowledge of the high molecular-mass forms of the enzyme. The use of monoclonal anti-GGT antibodies could be the starting point for new data concerning the multiple forms of the enzyme, especially those with high molecular mass: indeed, it is not unreasonable to imagine that these antibodies could be used in the assay of epitopes of serum GGT, assuming that some of them may be masked in specific pathophysiological conditions.

References

1 Artur, Y.: Détermination de la gamma-glutamyltransférase plasmatique et de ses formes multiples. Apport des techniques immunochimiques. Applications cliniques; Thèse de Doctorat d'Etat ès Sciences Pharmaceutiques, Nancy (1984).

2 Artur, Y.; Wellman-Bednawska, M.; Jacquier, A.; Siest, G.: Complexes of serum gamma-glutamyltransferase with apolipoproteins and immunoglobulin A. Clin. Chem. *30:* 631–633 (1984).

3 Artur, Y.; Wellman-Bednawska, M.; Jacquier, A.; Siest, G.: Associations between serum gamma-glutamyltransferase and apolipoproteins. Relationships with hepatobiliary diseases. Clin. Chem. *30:* 1318–1321 (1984).

4 Artur, Y.; Wellman-Bednawska, M.; Sanderink, G.J.; Siest, G.: Effect of sialic acid content on electro-immunodiffusion of human serum gamma-glutamyltransferase: clinical aspects. 5th Int. Congr. Clinical Enzymology, Jerusalem 1985.

5 Artur, Y.; Wellman-Bednawska, M.; Siest, G.: Immunological studies of human liver gamma-glutamyltransferases; in Siest, Heusghem, Gamma-glutamyltransferases. Advances in biochemical pharmacology; 3rd ser., pp. 61–67 (Masson, Paris, 1982).

6 Artur, Y.; Wellman-Bednawska, M.; Siest, G.: Formes moléculaires multiples de la gamma-glutamyltransférase plasmatique: caractérisation et signification clinique. Annls Biol. clin. *42:* 115–122 (1984).

7 Crofton, P.M.; Smith, A.F.: High-molecular-mass alkaline phosphatase in serum and bile. Nature and relationship with lipoprotein X. Clin. Chem. *27:* 867–874 (1981).

8 Hada, T.; Higashino, K.; Yamamoto, H.; Okochi, T.; Sumikawa, K.; Yamamura, Y.: Further investigations on a novel gamma-glutamyltranspeptidase in human renal carcinoma. Clinica chim. Acta *112:* 135–140 (1981).

9 Huseby, N.E.: Separation and characterization of human gamma-glutamyltransferases. Clinica chim. Acta *111:* 39–45 (1981).

10 Huseby, N.E.: Multiple forms of serum gamma-glutamyltransferases. Association of the enzyme with lipoproteins. Clinica chim. Acta *124:* 103–112 (1982).

11 Huseby, N.E.: Hydrophilic form of gamma-glutamyltransferase. Proteolytic formation in liver homogenates and its estimation in serum. Clinica chim. Acta *124:* 113–121 (1982).

12 Huseby, N.E.; Eide, T.J.: Variant gamma-glutamyltransferase in colorectal carcinoma and liver metastasis. A case study. Clin. Biochem. *18:* 217–219 (1985).

13 Izumi, M.: Isolation, characterization and clinical evaluation of the gamma-glutamyltransferase associated with hepatocellular carcinoma. Acta med. Okayama *39:* 19–33 (1985).

14 Kojima, J.; Kanatani, M.; Nakamura, N.; Kashiwagi, T.; Tohjoh, F.; Akiyama, M.: Electrophoretic fractionation of serum gamma-glutamyltranspeptidase in human hepatic cancer. Clinica chim. Acta *106:* 165–172 (1980).

15 Köttgen, E.; Gerok, W.: Isoenzymedifferenzierung der gamma-Glutamyltransferase mit Concanavalin A und Con-A Sepharose. Klin. Wschr. *54:* 439–444 (1976).

16 Masuike, M.; Ogawa, M.; Kitahara, T.; Murata, A.; Matsuda, K.; Kosaki, G.: Development of radioimmunoassay for gamma-glutamyltransferase using pancreatic enzyme. Ann. clin. Biochem. *20:* 247–250 (1983).

17 Miura, T.; Matsuda, Y.; Tsuji, A.; Katunuma, N.: Immunological cross-reactivity of gamma-glutamyltranspeptidases from human and rat kidney, liver and bile. J. Biochem. *89:* 217–222 (1981).
18 Scherberich, J.E.; Kleeman, B.; Mondorf, W.: Isolation of kidney brush border gamma-glutamyltranspeptidase from urine by specific antibody gel chromatography. Clinica chim. Acta *93:* 35–41 (1979).
19 Selvaraj, P.; Balasubramanian, K.A.: Soluble forms of gamma-glutamyltransferase in human adult liver, fetal liver, and primary hepatoma compared. Clin. Chem. *29:* 90–95 (1983).
20 Shaw, L.M.; Petersen-Archer, L.; London, J.W.; Marsh, E.: Electrophoretic, kinetic, and immunoinhibition properties of gamma-glutamyltransferase from various tissues compared. Clin. Chem. *26:* 1523–1527 (1980).
21 Sudo, K.; Kanno, T.: Membrane associated enzyme complexes containing lambda type of immunoglobulin A. Clinica chim. Acta *107:* 49–58 (1980).
22 Szewczuk, A.; Kuropatwa, M.; Lang, D.: Enzyme-linked immunosorbent assay (ELISA) and colorimetric determination of cow gamma-glutamyltransferase. Archs Immunol. ther. exp. *31:* 121–125 (1983).
23 Szewczuk, A.; Milnerowicz, H.; Sobiech, K.A.: Reaction of antibody with gamma-glutamyltranspeptidase. I. Isolation of the human enzyme and inhibition of its activity by antiserum. Archs Immunol. ther. exp. *25:* 589–600 (1977).
24 Taniguchi, N.; Yososawa, N.; Iizuka, S.; Sako, F.; Tsukada, Y.; Satoh, M.; Dempo, K.: Gamma-glutamyltranspeptidase of rat liver and hepatoma tissues: an enzyme immunoassay and immunostaining studies. Ann. N.Y. Acad. Sci. *417:* 203–212 (1983).
25 Tsuchida, S.; Imai, F.; Sato, K.: Immunological characterization of gamma-glutamyltransferases in human serum. J. Biochem. *89:* 775–782 (1981).
26 Wellman-Bednawska, M.; Artur, Y.; Siest, G.: Variations in sialic acid content of gamma-glutamyltransferase. A consequence for immunochemical determinations? Clinica chim. Acta *148:* 21–30 (1985).
27 Wenham, P.R.; Horn, D.B.; Smith, A.F.: Physical properties of gamma-glutamyltransferase in human serum. Clinica chim. Acta *141:* 205–218 (1984).
28 Wieland, H.; Bernhard, R.; Baggio, G.; Muller, P.; Seidel, D.: Demonstration of a lipoprotein-GGT complex in the plasma of cholestatic patients. Eur. J. clin. Invest. *7:* 243 (1977).
29 Yamamoto, H.; Sumikawa, K.; Hada, T.; Higashino, K.; Yamamura, Y.: Gamma-glutamyltransferase from human hepatoma tissue in comparison with normal liver enzyme. Clinica chim. Acta *111:* 229–237 (1981).

Y. Artur, PhD, Centre du Médicament, Laboratoire de Biochimie Pharmacologique, U.A. CNRS No. 597, 30, rue Lionnois, BP 403, F-54001 Nancy Cedex (France)

Adv. clin. Enzymol., vol. 3, pp. 90–94 (Karger, Basel 1986)

Creatine Kinase Variants

Current Status

Petter Urdal

Central Laboratory, Ullevål University Hospital, Oslo, Norway

In 1963, Burger first described the existence of three different forms of creatine kinase (CK). These variants, isoenzymes, were named CK-MM, CK-MB and CK-BB, names which indicate their subunit composition. During the last 10 years many other, additional, forms of CK have been described. We now know that all the three 'classical' isoenzymes may undergo postsynthetic changes. We also know that a fourth isoenzyme exists within the cell, and in contrast to the cytoplasmic forms mentioned above is attached to the mitochondrial membrane. The different CK variants now known to occur are summarized in table I and their approximate electrophoretic mobilities are shown in figure 1.

CK-MM

CK-MM is relatively organ specific, being present only in skeletal muscle and myocardium [1], and is released following damage to these organs.

A recent report by Guslits and Jacobs [2] suggests that several different forms of CK-MM are present in the cytoplasm of the cell, but whether this represents real synthetic heterogeneity, differences introduced during the synthesis of CK-MM, or postsynthetic heterogeneity, or possibly both, is not yet known. However, the various forms do not appear to differ very much in structure. Thus, they all have a rather similar electrophoretic mobility and their activities are completely inhibited by CK-M subunit inhibitory antibodies.

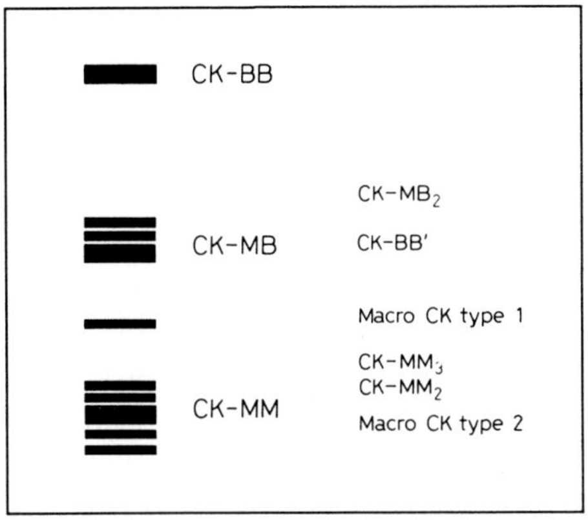

Fig. 1. Electrophoretic mobility of CK variants.

Table I. Different CK variants

A Cytoplasmic isoenzymes	
CK-MM	
CK-MB	
CK-BB	
Post-synthetically modified variants occurring in blood:	
CK-MM and CK-MB:	with changed electrophoretic mobility
CK-BB:	with changed electrophoretic mobility
	inactivated through binding to IgG
	bound to high-affinity IgG (macro CK type 1)
B Mitochondrial isoenzyme	
CK-MiMi	
Various aggregate forms in blood (macro CK type 2)	

Wevers et al. [3] first observed that when CK-MM enters the circulation following tissue damage, there is a time-dependent change in its electrophoretic mobility. Further work by Wevers, Chapelle and Heusghem [4] and others [5] has shown that a thermolabile factor, most likely an enzyme, is responsible for this conversion. Probably, the conversion is brought about by the splitting off of a few amino acids. The remaining part of CK-

MM moves slightly more anodally on agarose gel electrophoresis but appears unchanged immunologically. Both Wevers and Chapelle give further details on this interesting phenomenon in this volume.

CK-MB

CK-MB is present mainly in the myocardium but also to a minor extent in skeletal muscle and is released following myocardial damage (acute myocardial infarction) and extensive skeletal muscle damage.

CK-MB also changes its electrophoretic mobility when entering blood, reflecting that the thermolabile factor mentioned above appears to attack the M subunit of CK, irrespective of whether it is linked to an M subunit (CK-MM) or a B subunit (CK-MB).

CK-BB

CK-BB is present in substantial concentrations in several organs including brain, stomach, colon, uterus and prostate gland [1], and therefore is by no means brain specific. Probably due to this wide tissue distribution, increased CK-BB in blood has been observed in many different clinical conditions [6, 7]. However, major increases in blood CK-BB do occur mostly in patients with brain damage: in fact in some conditions increased CK-BB now appears to reliably and usefully indicate brain damage [8], and in some patients with cancer, especially when of prostatic or pulmonary origin.

Several variants of CK-BB have been described (table I). Cho et al. [9] found that tissue CK-BB added to serum changed its electrophoretic mobility during incubation at 37 °C. After 3 h of incubation an MB-like band occurred and after 17–42 h a second new band, migrating between MM and MB, was also present. This CK-BB of altered mobility appeared to have an unchanged molecular weight. At present the molecular basis for the changes observed is not known.

Nealon and Henderson [10] reported another change that took place when purified CK-BB was incubated at 37 °C in the presence of either serum or purified IgG, i.e. the formation of a complex between the enzyme and IgG. The enzymatic activity of the complex was extremely labile towards inactivation and was in this respect clearly different from the BB activity of the macro CK-BB (macro CK type 1) complex, discussed below. These complexes probably have an electrophoretic mobility different from that of CK-BB, but the complex, due to its lability, is probably not observable in an electrophoretic system.

Some sera (0.5–2.0% of all sera) contain CK-BB linked to IgG through a high-affinity bond [11–13] ('macro CK type 1'). This IgG is remarkably specific for CK-BB and binds the enzyme, through its Fab region, even in the presence of excess amounts of CK-MM and CK-MB. The IgG thus appears to be an autoantibody directed against CK-BB [13], but at present it is not known why some persons, mostly elderly women, produce this immunoglobulin. Further details on macro CK are given by Stein and coworkers in this volume.

Mitochondrial CK

CK-mit is probably composed of two identical subunits [14] and is present in skeletal muscle, myocardium, liver and brain [14–16], possibly also in several other organs. The CK-mit found in these different organs are very similar, possibly identical. Thus, antibodies produced against heart CK-mit also react with liver CK-mit [15].

Release of CK-mit into blood is mostly seen in patients who are severely ill (cancer, liver cirrhosis) [17]. In blood, CK-mit is present in different aggregate forms, either of increased molecular weight ('macro CK type 2') or of normal size (M_r 80,000) [15]. These variants move cathodically to CK-MM on electrophoresis. The high molecular form may be reduced in vitro to normal size form by incubation with urea [15].

These CK variants have attracted considerable interest up to now, and will still do so for two reasons. Firstly, it will be of interest to know more about these particular variants, their way of production, characteristics and clinical significance, and, secondly, when present but not correctly identified some of these variants may be misinterpreted to be of CK-MB nature when serum is analysed for its content of different CK isoenzymes.

References

1 Urdal, P.; Urdal, K.; Strømme, J.H.: Cytoplasmic creatine kinase isoenzymes quantitated in tissue specimens obtained at surgery. Clin. Chem. *29:* 310–313 (1983).

2 Guslits, B.G.; Jacobs, H.K.: Investigation of the heterogeneity of the MM isoenzyme of creatine kinase. Clinica chim. Acta *130:* 55–69 (1983).

3 Wevers, R.A.; Delsing, M.; Klein Gebbink, J.A.G.; Soons, J.B.J.: Post-synthetic changes in creatine kinase isozymes (EC 2.7.3.2). Clinica chim Acta *86:* 323–327 (1978).
4 Chapelle, J.P.; Heusghem, C.: Further heterogeneity demonstrated for serum creatine kinase isoenzyme MM. Clin. Chem. *26:* 457–462 (1980).
5 Lott, J.A.; Heinz, J.W.: Transformation of creatine kinase-BB isoenzyme in vitro. Effect of carboxylic acids, thiols, pH, cations and chelators. Clin. Chem. *28:* 2414–2417 (1982).
6 Tsung, S.H.: Several conditions causing elevation of CK-MB and CK-BB. Am. J. clin. Path. *75:* 711–715 (1981).
7 Coolen, R.B.; Pragay, D.A.; Nosanchuk, J.S.; Belding, R.: Elevation of brain-type creatine kinase in serum from patients with carcinoma. Cancer *44:* 1414–1418 (1979).
8 Urdal, P.; Strømme, J.H.: Creatine kinase BB in cerebrospinal fluid and blood: methodology and possible clinical application (Editorial review). Scand. J. clin. Lab. Invest. *45:* 481–487 (1985).
9 Cho, H.W.; Meltzer, H.Y.; Joung, J.; Goode, D.: Effect of incubation in human plasma on electrophoretic mobility of brain-type creatine phosphokinase. Clinica chim. Acta *73:* 257–265 (1976).
10 Nealon, D.A.; Henderson, A.R.: Creatine kinase-1 is principally inactivated in serum by complexing with immunoglobulin-G. Clin. Sci. *58:* 157–160 (1980).
11 Urdal, P.; Landaas, S.: Macro creatine kinase BB and some data on its prevalence. Clin. Chem. *25:* 461–465 (1979).
12 Bohner, J.; Stein, W.; Steinhart, R.; Eggstein, M.: Serum creatine kinase BB linked to immunoglobulin G. Clinica chim. Acta *97:* 83–88 (1979).
13 Urdal, P.: Macro creatine kinase BB isoenzyme in serum. Most likely an antigen-autoantibody complex. Scand. J. clin. Lab. Invest. *41:* 499–505 (1981).
14 Wevers, R.A.; Mul-Steinbusch, M.W.F.J.; Soons, J.B.J.: Mitochondrial CK (EC 2.7.3.2) in the human heart. Clinica chim. Acta *101:* 103–111 (1980).
15 Kanemitsu, F.; Kawanishi, I.; Mizushima, J.: Characteristics of mitochondrial creatine kinases from normal human heart and liver tissues. Clinica chim. Acta *119:* 307–317 (1982).
16 Wevers, R.A.; Reutelingsperger, C.P.M.; Dam, B.; Soons, J.B.J.: Mitochondrial creatine kinase (EC 2.7.3.2) in the brain. Clinica chim. Acta *119:* 209–223 (1981).
17 Wu, A.H.B.; Herson, V.C.; Bowers, G.N., Jr.: Macro creatine kinase types 1 and 2. Clinical significance in neonates and children as compared with adults. Clin. Chem. *29:* 201–204 (1983).

P. Urdal, MD, University of Oslo, Central Laboratory, Ullevål Hospital,
N–0407 Oslo 4 (Norway)

Adv. clin. Enzymol., vol. 3, pp. 95–104 (Karger, Basel 1986)

Macro Creatine Kinases[1]

W. Stein, J. Bohner, M. Bahlinger

Medizinische Universitätsklinik, Abteilung 4, Tübingen, BRD

Creatine kinase (CK; EC 2.7.3.2) may be present in human serum not only as the well-known cytoplasmic isoenzymes CK-MM, CK-MB and CK-BB, but also as so-called CK variants [1]. In 1978, Yuu et al. [2] first used the term 'macro creatine kinase' for a CK with a relative molecular mass (M_r) of about 325,000 found in the serum of a patient suffering from breast cancer. In recent years, a variety of CK variants have been described [1–5] which have gained increasing interest in clinical chemistry laboratories, primarily because these variants interfere with laboratory diagnosis of acute myocardial infarction (AMI) based on determinations of CK-MB activities.

During our prospective study comprising about 3,000 patients [5], we were astonished to see that in our clinic, typical of university hospitals for internal medicine, there is no such variety in CK patterns as might be deduced from the literature. Besides the ordinary cytosolic isoenzymes CK-MM, CK-MB and CK-BB, respectively, we detected 126 cases with atypical CK activities. Without exception these 126 variant CK showed increased M_r of $> 200,000$ and could be differentiated into only two different classes of macro CK, which we have named macro CK type 1 and macro CK type 2 [6–8]. Whereas macro CK type 1 is recognized as an immune complex formed between immunoglobulins (Ig) and CK isoenzymes [9, 10], macro CK type 2 is no such immune complex [5]. Biochemical data and clinical observations suggest that macro CK type 2 is of mitochondrial origin.

[1] This study was supported by the Deutsche Forschungsgemeinschaft.

Detection and Differentiation of Macro CK

A reliable assay for macro CK must be based on the proof of an increased M_r. This can be done in one out of two possible ways. Firstly, a direct confirmation of the increased M_r may be obtained from exclusion chromatography or polyacrylamide-gradient gel electrophoresis [1]. Secondly, an increase of M_r may be assumed if results of immunological procedures or reaction with protein A indirectly indicate that the CK moiety is linked to another serum protein. The methods commonly applied in the clinical laboratory for determination of CK isoenzymes (isoenzyme electrophoresis, immunoinhibition, ion-exchange chromatography), however, cannot unambiguously demonstrate the presence of macro CK: they only may be used as screening methods to detect macro CK-suspected cases of which the exact nature must be elucidated by more specific methods. For differentiation of both types of macro CK, we obtained best results from using immunoinhibition of CK-M subunits in combination with either isoenzyme electrophoresis or determination of the activation energy [5].

We now report our results from 141 cases of macro CK type 1 and of 105 cases of macro CK type 2 which we detected during our prospective study or which were sent to our laboratory during recent years.

Biochemical Properties of Macro CK Type 1

The term macro CK type 1 comprises the immunoglobulin-bound CK. Whereas the occurrence and composition of macro CK-BB now is well documented [9, 10], we obtained macro CK-MB only in vitro after incubation of human CK-MB with high affinity anti-CK-BB autoantibodies isolated from a macro CK-BB serum. These macro MB complexes, however, exhibited a markedly lower binding affinity; therefore, we assume a cross-reaction between the anti-BB antibodies and the B subunit of CK-MB [9]. Remarkably, we never detected CK-MM binding with immunoglobulins [8].

In 102 cases we exclusively found CK-BB linked to either IgG (n = 96) or IgA (n = 2) or both (n = 4). The ratio of κ:λ light chains was 5:1. The binding between CK-BB and Ig fulfills the criteria which are typical of specific antigen-antibody reactions:

(1) Ig shows a limited binding capacity for CK-BB; its concentration in serum usually exceeds that of CK-BB.

(2) The binding is reversible: after acidolytic dissociation of the complexes and chromatographic purification of the Ig fraction, recombination of macro CK-BB is achieved with autologous or homologous CK-BB.

(3) The binding is specific for catalytically active CK-BB: there is only a small cross-reaction with CK-MB, but no reaction with CK-MM or mitochondrial CK. CK-BB is linked to the Fab region of IgG with a high affinity of 10^{11} liters/mol [10] resulting in 2:1 complexes between CK-BB and IgG (M_r: 330,000) [9].

The typical pattern after electrophoresis is shown in figure 1: macro CK type 1 mostly migrates anodally to the origin, showing one or two bands between CK-MM and CK-MB, sometimes co-migrating with these cytosolic isoenzymes.

In contrast to free, normal-size CK-BB, macro CK-BB shows a markedly higher stability towards shifts of the pH value and especially towards thermal inactivation (table I). We now recommend to utilize this increased thermal stability of macro CK for the differentiation of unexpected results during isoenzyme determinations of CK [11].

Further typical biochemical data of macro CK type 1 are summarized in table I.

Biochemical Properties of Macro CK Type 2

During exclusion chromatography and polyacrylamide-gradient gel electrophoresis we observed characteristic activity patterns corresponding to $M_r > 750{,}000$ in 5 of 57 cases, to $M_r \approx 300{,}000$ in 22/57 cases and to both classes of M_r in 30/57 cases. The occurrence of these patterns of various molecular masses of macro CK type 2 agrees well with our results of isoenzyme electrophoresis, which showed at least one and mostly two out of three possible atypical bands (fig. 1).

After heat inactivation, macro CK type 2, like macro CK type 1, shows a remarkably high residual activity (table I), which at pH 8.0 is comparable to the stability of CK-MM. Only 18 samples containing considerable amounts of heat-labile, normal-size CK-BB in addition to macro CK type 2 showed a residual CK activity of less than 0.4.

Macro CK type 2 demonstrates a high dependence of its catalytic activity on temperature. To assess this dependence of enzyme activity on temperature, we determined CK activity at 25, 30 and 37 °C. The slope of the Arrhenius plots (plot of ln (U/l) vs. 1/T (1/K)) is used to calculate the

Table I. Characteristics of macro CK

Methods	Type 1 (Ig-linked BB)	Type 2 (serum form of mitochondrial CK)
1 Relative molecular mass [1]	330,000	≈ 300,000 > 750,000
2 Ion-exchange chromatography [5, 6]	elutes at	
DEAE-Sepharose	140 mmol/l Na^+	100 mmol/l Na^+
CM-Sepharose	void volume	100 mmol/l Na^+
3 Immunological reactivity [1, 5, 6, 8, 9]		
With anti-CK-B	yes	no
With anti-CK-M	no	no
With anti-Ig	yes	no
With protein A	yes (IgG-linked CK)	no
4 Incorporation of ^{125}I-labeled [8]		
CK-BB	yes	no
CK-MM	no	no
5 Catalytic activity [1, 5–7, 11]		
Stability (37 °C, pH 7.5), half-life	> 50 h (CK-MB: ≈ 4 h)	≈ 12 h
Michaelis constants (30 °C)		
ADP	0.07 mmol/l	0.06 mmol/l
Creatine phosphate	1.6 mmol/l	0.5 mmol/l
pH optimum	6.0	7.3
Activation energy	≈ 50 kJ/mol	≈ 120 kJ/mol (CK_{mito}:130 kJ/mol)

activation energy Ea [7]. Besides thermostability, a high activation energy is particularly characteristic of macro CK type 2 [5, 7], and has never been associated with normal-size, dimeric, cytoplasmic isoenzymes of CK. Its apparent Ea of about 114 kJ/mol significantly exceeds those of the cytoplasmic isoenzymes: CK-BB: 62 kJ/mol [1]; CK-MB: 58 kJ/mol [1]; CK-MM 1: 67 kJ/mol [12]; CK-MM 2: 59 kJ/mol [12]; CK-MM 3: 49 kJ/mol [12], and of macro CK type 1 (table I).

Further typical biochemical properties of macro CK type 2 are summarized in table I.

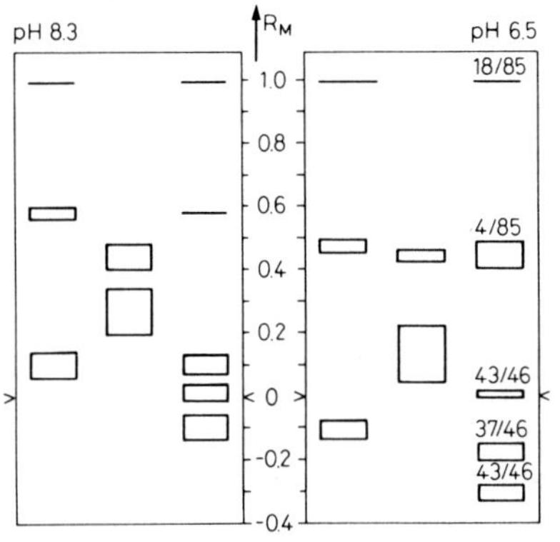

Fig. 1. Electrophoretic patterns of CK forms. Relative mobilities (R_M) were calculated with reference to CK-BB ($R_M = 1.0$) and the point of application ($R_M = 0.0$), indicated by arrows. The areas of the bands correspond to average values (± SD) for R_M. The fractions (e.g. 18/85) indicate that this particular band was seen in 18 of 85 cases. The components in each lane are as follows: For runs at pH 8.3 (lanes from left to right): lane 1: CK-MM (0.10), CK-MB (0.58), CK-BB (1.0); lane 2: IgG-linked CK-BB (0.27), IgA-linked CK-BB (0.44); lane 3: macro CK type 2 (−0.10, +0.01, 0.10), CK-BB′ = pseudo-MB (0.58), CK-BB (1.0). For runs at pH 6.5: lane 1: CK-MM (−0.11), CK-MB (+0.47), CK-BB (1.0); lane 2: IgG-linked CK-BB (0.13), IgA-linked CK-BB (0.44); lane 3: macro CK type 2 (−0.31, −0.18, 0.0), CK-BB′ (0.44), CK-BB.

Our results of physicochemical, biochemical, and immunological studies showed that cytosolic isoenzymes are not involved in the formation of macro CK type 2. Furthermore, we could exclude the possibility that this type of macro CK is an immune complex or that lipoproteins are an essential part of macro CK type 2 [5]. On the other hand, comparison of serum macro CK type 2 with mitochondrial CK isolated from human tissues revealed that the properties of both CK are in agreement [5]. We observed: (1) similar physicochemical properties: isoelectric points, relative mobilities during electrophoresis, formation of high-molecular mass complexes; (2) comparable enzyme kinetics: heat stability, low Michaelis constants (K_m), extraordinarily high activation energy, and (3) common immunological properties and no reactions with anti-CK-M, anti-CK-B, and anti-Ig antibodies as well as protein A.

Though a final proof of the identity of both these CK forms by use of, for example, anti-mitochondrial CK antibodies is still lacking, we assume that macro CK type 2 is the manifestation of mitochondrial CK in serum.

Macro CK in Hospitalized Patients

Most patients with macro CK type 1 (95%) and most patients with macro CK type 2 (88%) – as detected in our prospective study – showed normal values for total CK activity. Residual activity after immunoinhibition of the CK-M subunit was often increased and sometimes exceeded 50 U/l (25 °C). Therefore, both types of macro CK are characterized by a relatively high percentage of residual CK activity after immunoinhibition of the CK-M subunits: the ratio of residual CK after immunoinhibition (calculated as 'CK-MB') to total CK activity very often exceeds 0.25. Not more than 5 (4%) patients showed both elevated values for total CK and a ratio between 0.06 and 0.25, a constellation which otherwise is suggestive of myocardial infarction. In our patients with suspect of AMI possible interference by macro CK was less than 0.4%, if we applied a stepwise diagnostic strategy as outlined below. In contrast to AMI, however, the enzyme patterns in patients with elevated macro CK in blood tend to remain fairly constant during several hours; we therefore do not observe dynamic activity-time curves. This persistence of CK activity is a further characteristic property of macro CKs, and is most significantly seen in patients with macro CK type 1. We were able to document persistent activities of IgG-linked CK-BB in 2 patients for more than 7 years. The occurrence of CK-BB-binding immunoglobulins in the blood leads to an accumulation of catalytically active CK-BB, because the CK-BB molecules, which normally are present only at very low concentrations, are now linked to these antibodies. Furthermore, the formation of the immune complex stabilizes the CK-BB activity in the blood and also decreases the rate of elimination of the CK-BB. Consequently, the level of the macro CK-BB activity is primarily under the control of the concentration of the CK-BB-binding autoantibodies and not of the amount of CK-BB which is released normally or after damage of a CK-BB-containing tissue. This may further explain why patients with macro CK type 1 show no common diagnosis apart from

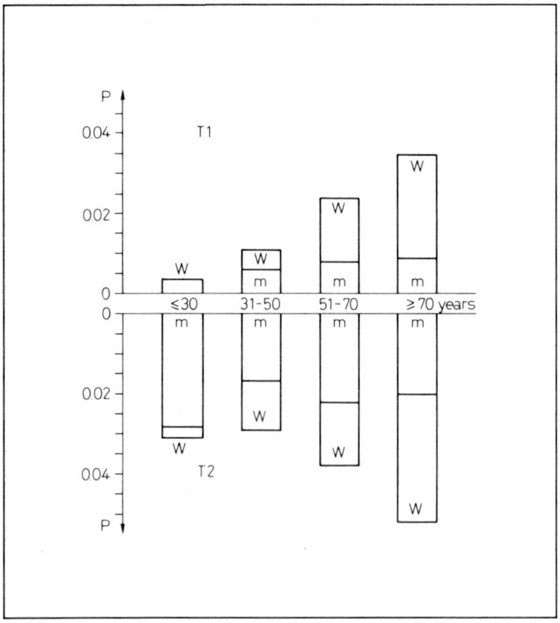

Fig. 2. Prevalence (P) of macro CK. Dependence on sex and age. Upper panel: Results for macro CK type 1 (T1), n = 36 hospitalized patients. m = Men; w = women. Lower panel: Results for macro CK type 2 (T2), n = 75 hospitalized patients.

diseases of the cardiovascular system which apparently only reflects the fact that macro CK type 1 is more often seen in older people (fig. 2). In contrast, activity of macro CK type 2 slowly parallels the stage of a severe disease, remains present until the patient's death or, during convalescence, slowly decreases and finally may disappear totally.

Patients with macro CK type 2 primarily suffer from malignancies (40%) and liver cirrhosis (20%), but we also observed it in serious acute diseases such as toxic epidermal necrosis and in children. These observations are in perfect agreement with the assumption that macro CK type 2 is released from the mitochondria after severe cell damage, during high rates of cell proliferation or production by a tumour.

The frequency with which macro CK is detected depends most critically on the sensitivity of the methods applied and the population studied. The more sensitive the method and the lower the discrimination limits and the older the patients and the more severely ill they are, the more often is macro CK seen. The prevalence of macro CK in all patients in our hospital ranges from less than 0.2 to 1.8 (type 1) and 3.7% (type 2), dependent on the discriminator chosen. In out-patients we determined a highest prevalence of macro CK type 2 of 1.1%; this is in agreement with a lower percentage of seriously ill patients in this group.

The dependence of macro CK on age and sex of the patients is documented in figure 2, demonstrating highest prevalences of macro CK in the group of women older than 70 years.

Interference of Macro CK during Diagnosis of Myocardial Infarction

Three types of interference by macro CK in patients' sera are known:

(1) Direct interference of the whole macro CK molecule in methods not absolutely discriminating CK-MB or CK-MM from macro CK, and CK-MB from 'pseudo-MB' or CK-BB′ (fig. 1) [5]: co-migration or co-elution of bands during electrophoresis [8] or chromatography [13] results in a misinterpretation of the 'MB' band or in an erroneously calculated ratio of CK-MB to CK-MM. This problem may be overcome by running the samples with and without inhibiting antibodies to CK-M to clearly locate CK-MB and CK-MM [8].

(2) Interference by the catalytically active CK moiety in some immunological methods. Using the immunoinhibition procedure, the coincidence of macro CK and myocardial infarction is recognized by the typical increase, peak, and decrease of total CK and residual CK after inhibition, while the ratio of CK after immunoinhibition to total CK exceeds 0.25 and shows an inverse course. In contrast to typical patterns of AMI, here CK after immunoinhibition and the ratio will not return to normal after a few days but will remain at their former, persistently high levels.

(3) Interference by the CK-BB-binding autoantibodies which are present in serum in excess relative to CK-B and which, due to a certain cross-reaction with CK-MB, may 'mask' CK-MB in electrophoresis [14] and immunoassays. The influence of macro CK type 1 on three different immunological assays for CK-MB is demonstrated in figure 3. The CK-MB result reported by each of the three solid-phase enzyme-immunoassays was negatively biased if CK-B-binding autoantibodies were present in the sample. Consequently, the influence of these autoantibodies may cause an apparently delayed – or, in special cases of small infarctions, missing – increase of CK-MB after AMI.

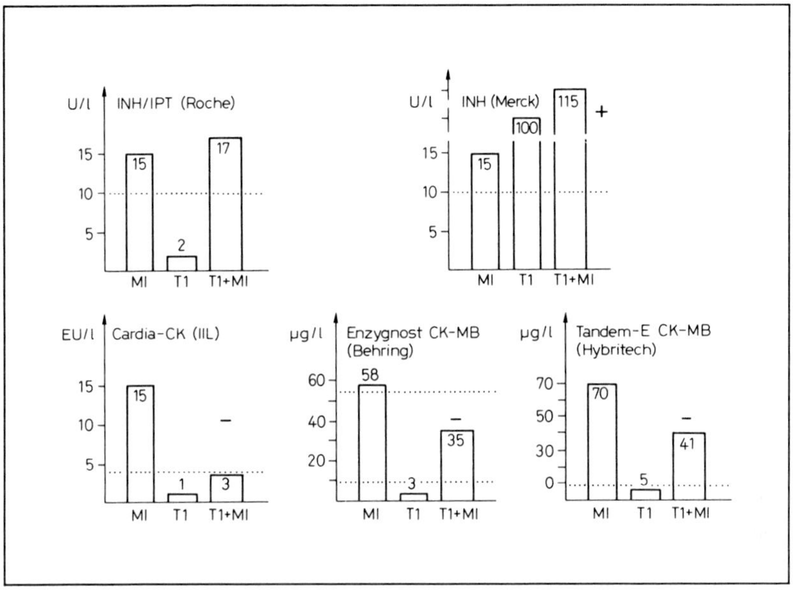

Fig. 3. Influence of macro CK type 1 on assays for CK-MB. Five assays (INH/IPT: Isomune-CK, Roche; INH: Immunoinhibition, Merck; Cardia-CK, IIL; Enzygnost CK-MB, Behring; Tandem-E CK-MB, Hybritech) were performed according to the manufacturers' instructions. Column MI = CK-MB result for the serum of a patient with infarction (serum diluted 1:1 with a normal serum). Column T1 = CK-MB result for the macro CK type 1 serum (1:1 diluted with normal serum). Column T1 + MI = 'CK-MB' result for the 1:1 mixture of the MI and the T1 serum. The dotted line indicates upper reference limits. INH/IPT: No interference, results as expected: very low activity in the T1 serum, and column T1 + MI equals the sum of column MI and column T1. INH: Positive interference (+) in the T1 serum, result of column T1 + MI nevertheless equals the sum of column MI and column T1 as expected. IIL, Behring, Hybritech: Low values for the T1 serum as expected. The surprisingly high negative interference (–) in the values of column T1 + MI indicates 'masking' of CK-MB by the auto-antibodies of the T1 serum.

Accordingly, we conclude that, regardless of the methods used for CK-MB, early exclusion and confirmation of AMI still requires a strictly defined diagnostic strategy [15] based on: (1) determination of CK-MB only in samples from patients suspected of AMI with elevated values for total CK activity; (2) sequential sampling during the diagnostic 'time window' within the first 20 h after onset of symptoms, and (3) observation of the typical enzyme pattern over time, because performing only one CK/CK-MB determination will give insufficient diagnostic information.

References

1 German Society for Clinical Chemistry: Report on the workshop conference on creatine kinase variants. J. clin. Chem. clin. Biochem. *21:* 859–876 (1983).

2 Yuu, H.; Takagi, Y.; Senju, O.; Hosoya, J.; Gomi, K.; Ishii, T.: Creatine kinase isoenzyme of high relative molecular mass in serum of a cancer patient. Clin. Chem. *24:* 2054–2057 (1978).

3 Lang, H.; Würzburg, U.: Creatine kinase, an enzyme of many forms (Review). Clin. Chem. *28:* 1439–1447 (1982).

4 Kwong, T.; Arvan, D.: How many creatine kinase 'isoenzymes' are there and what is their clinical significance (Review)? Clinica chim. Acta *115:* 3–8 (1981).

5 Stein, W.; Bohner, J.; Bahlinger, M.: Analytical patterns and biochemical properties of macro creatine kinase type 2. Clin. Chem. *31:* 1952–1958 (1985).

6 Barnett, R.; Goetz, W.: CK-MB, methods and clinical significance. Proc. CK-MB symposium, Philadelphia 1981, pp. 61–73 (GIT Verlag, Darmstadt 1981).

7 Stein, W.; Bohner, J.; Steinhart, R.; Eggstein, M.: Macro creatine kinase: determination and differentiation of two types by their activation energies. Clin. Chem. *28:* 19–24 (1982).

8 Bohner, J.; Stein, W.; Steinhart, R.; Würzburg, U.; Eggstein, M.: Macro creatine kinases: results of isoenzyme electrophoresis and differentiation of the immunoglobulin-bound type by radioassay. Clin. Chem. *28:* 618–623 (1982).

9 Stein, W.; Bohner, J.; Krais, J.; Müller, M.; Steinhart, R.; Eggstein, M.: Macro creatine kinase BB: evidence for specific binding between creatine kinase BB and immunoglobulin G. J. clin. Chem. clin. Biochem. *19:* 925–930 (1981).

10 Urdal, P.: Macro creatine kinase BB isoenzyme in serum: most likely an antigen-autoantibody complex. Scand. J. clin. Lab. Invest. *41:* 499–505 (1981).

11 Bohner, J.; Stein, W.; Renn, W.; Steinhart, R.; Eggstein, M.: Stability of macro creatine kinases and creatine kinase isoenzymes compared: heat inactivation test for determination of thermostable creatine kinases. J. clin. Chem. clin. Biochem. *19:* 1021–1026 (1981).

12 Wevers, R.; Hagelauer, U.; Stein, W.; Bohner, J.; Faust, U.; Landeghem, A. van; Soons, J.: Indices for the age of the creatine kinase M-chain in the blood. Clinica chim. Acta *148:* 197–210 (1985).

13 Stein, W.; Bohner, J.; Eggstein, M.; Lang, H.: Eigenschaften und Bedeutung von Makro-Kreatinkinasen. J. clin. Chem. clin. Biochem. *17:* 739–745 (1979).

14 Pudek, M.; Jacobson, B.: Falsely negative laboratory diagnosis for myocardial infarction owing to the concurrent presence of macro creatine kinase and macro lactate dehydrogenase. Clin. Chem. *28:* 2434–2437 (1982).

15 Gerhardt, W.; Waldenström, J.; Hörder, M.; Hovfendahl, S.; Billström, R.; Ljungdahl, R.; Berning, H.; Bagger, P.: Creatine kinase and creatine kinase B-subunit activity in cases of suspected myocardial infarction. Clin. Chem. *28:* 277–283 (1982).

W. Stein, MD, PhD, Medizinische Universitätsklinik, Abteilung 4,
D–7400 Tübingen (FRG)

Adv. clin. Enzymol., vol. 3, pp. 105–115 (Karger, Basel 1986)

Clinical Application of the Heterogeneity of Creatine Kinase Isoenzyme MM

Jean-Paul Chapelle

Department of Clinical Chemistry, University of Liège, Belgium

Creatine kinase (CK, EC 2.7.3.2) exists in human tissues as a dimer composed of two subunits, designated M and B [1]. These subunits combine to produce three isoenzymes, termed MM, MB and BB. Within the last few years, however, the presence of different forms of CK in the blood, other than the three classical isoenzymes, has been reported: these atypical forms differ from MM, MB and BB in their behaviour on electrophoresis and ion-exchange chromatography [2, 3].

Independent of the atypical forms of CK, both MM and MB can be shown to be heterogeneous; several subisoenzymes – or isoforms – can be identified when the samples are submitted to separation techniques providing high resolutions. In 1972, Smith [4] observed that the MM isoenzyme could be split into three distinct subbands using polyacrylamide-gel electrophoresis; the presence of three isoforms in serum specimens was confirmed by Wevers et al. [5, 6] by prolonging the usual time for standard electrophoresis on agarose gel. Using thin-layer isoelectric focusing [7], we determined the isoelectric points (pI) of these three forms as 6.88 (MM_1), 6.70 (MM_2) and 6.45 (MM_3). Recently, we characterized three additional MM isoforms [8], with pIs 7.29, 7.10 (MM_0) and 6.25 (MM_4), while other reports suggested that the MM heterogeneity could be even more important [9, 10]. As to the MB isoenzyme, it appeared to consist of two isoforms, MB_1 and MB_2, with pIs of 5.61 and 5.34, respectively [7].

In patients with acute myocardial infarction (AMI) and in subjects who underwent coronary bypass surgery, we observed that the MM and MB isoform patterns changed with time after the enzyme release [7, 8]. The first isoforms to appear in the blood, MM_0 and MM_1, regularly decreased after AMI, whereas MM_2, MM_3 and MM_4 progressively increased; MM_3, for example, became the prominent MM isoform as early as 12 h after serum CK reached its peak. The transformation of human CK-MM_1 to MM_2 and MM_3 was confirmed by several authors [11–13]. We also showed that the MM_0 and MM_1 isoforms were in approximately equal proportions in human myocardium specimens taken within 5 h after death, representing together nearly 85% of the total MM isoenzyme; most of the remaining activity was due to the isoform of pI 7.29, designated throughout the present study as MM_{-1}. When myocardium homogenates were incubated with normal human serum, the labile cellular isoform MM_{-1} was totally inactivated within a few hours; MM_0 rapidly decreased, fell below 2% of the total MM fraction after 4 h and could no longer be demonstrated after 12 h. The MM_2, MM_3 and MM_4 isoforms successively appeared and, after 20 h, represented 90% of the MM activity, confirming the observations made after AMI. Moreover, when purified MM_0 was incubated with normal serum, the MM_1 subband appeared within 15 min and its formation always preceded that of MM_2 and MM_3; conversely, MM_0 never arose from MM_1 or from any other MM subbands [8].

As heat inactivation of normal serum prior to incubation with myocardium extracts, or addition of EDTA to a final concentration of 5 mmol/l inhibited the transformation of the MM cellular forms [14], we reached the conclusion that the postsynthetic changes recorded for CK were due to a thermolabile, probably calcium-dependent constituent of human serum. Falter et al. [12, 15] partly purified the 'CK conversion factor' and determined its molecular weight. Recently, several authors [16–18] demonstrated that the serum carboxypeptidases were responsible for the progressive removal of the two C-terminal lysines from each of the M subunits of MM_1, successively producing MM_2 and MM_3 (fig. 1). The other steps of the degradation process, namely the mechanism leading to the transformation of MM_0 to MM_1, or to the formation of MM_4, are still to be elucidated.

Important progress in the management of AMI patients may result from a better knowledge of enzyme release from the jeopardized myocardium. In the present study, we investigated the changes occurring in the native MM isoforms during the first hours after their liberation into the blood. Therefore, we serially measured the serum MM_{-1}, MM_0 and MM_1

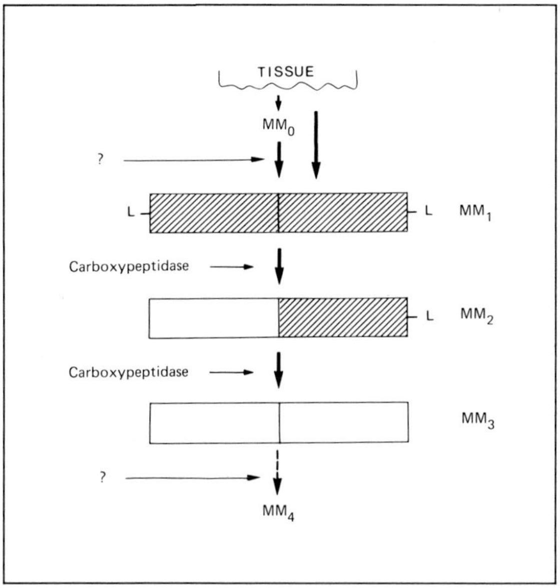

Fig. 1. Degradation process of the tissue CK isoforms MM_0 and MM_1 after their release into the blood. L = Lysine.

levels after endomyocardial biopsy and in the acute phase of myocardial infarction. In patients with AMI, the effect of an increased washing-out of the necrotic area (thrombolytic therapy) on the serum MM isoform patterns was also studied.

Material and Methods

Patient Population

Our study population consisted of:

58 AMI patients (48 men and 10 women, mean age 57, SD 10 years) who were admitted to the Coronary Care Unit (CCU), University Hospital, Liège, Belgium. AMI was diagnosed on the basis of a typical clinical history, electrocardiographic evidence, and a characteristic increase and decrease of total CK activity in serum. To be included in the study, patients had to reach the CCU within 4 h following the onset of chest pain and initial CK values had to be < 150 U/l. Thirty-three patients received a fibrinolytic therapy (intravenous perfusion of 0.8–1.0 mega-units of streptokinase within 30 min) within 15 min following admission, whereas the other 25 patients were treated in a conventional manner (administration of continuous intravenous heparin, 1,000 U/h). For all patients, blood was sampled upon admission, every 4 h during the first 36 h, and every 12 h thereafter until 72 h.

17 patients (13 men and four women, mean age 49, SD 12 years) who underwent *elective endomyocardial biopsy*. Endomyocardial biopsy was carried out in the right ventricular cavity and was associated with a right heart catheterization (Swan-Ganz catheter), using a percutaneous right femoral vein approach. A King bioptome was used, and 5–6 samples were taken per patient, according to a technique previously described [19]. Two-dimensional echocardiography was performed using a Diasonics V 3400 R apparatus with a 2.25-MHz transducer. There was no concomitant coronary angiography; intramuscular premedication was not given. The most frequent diagnosis in our series was myocarditis.

The initial venous blood sample was obtained 30 min before initiation of the procedure and the second sample at the end of biopsy (on average, 30 min later). Measurements were also performed after 2, 4, 6, 8, 12, 16, 20 and 24 h.

CK-MM Isoform Pattern Determination

Serum samples were stored at 4 °C and analysed within 24 h for the CK-MM isoforms. EDTA in a final concentration of 5 mmol/l was added to the samples in order to avoid in vitro transformations of the labile isoforms [14]. CK-MM isoforms were determined after separation by thin-layer isoelectric focusing on polyacrylamide gel, with specific CK staining and densitometric scanning as previously described [7].

Total CK and Lactate Dehydrogenase (LD) Activity Measurements

Total CK (EC 2.7.3.2) and LD (EC 1.1.1.27) activities were measured by optimized spectrophotometric methods [20, 21] (CK and LDH UV tests, No. 3388 and 3399, Merck, Darmstadt, FRG) on a Bichromatic Analyzer ABA 100 (Abbott Lab., North Chicago, Ill.). Reference intervals were 0–100 U/l for total CK and 150–350 U/l for LD.

Table I. Evolution of the serum CK-MM isoforms in 25 AMI patients with short admission delays (treatment: intravenous heparin)

MM isoforms % of total CK-MM activity	Time after hospital admission												
	0 h	4 h	8 h	12 h	16 h	20 h	24 h	28 h	32 h	36 h	48 h	60 h	72 h
MM_{-1}	0.0	0.3	0.1	0.1	0.2	0.2	0.1	0.2	0.2	0.0	0.0	0.0	0.0
MM_0	6.7	11.9	10.9	9.9	7.6	5.6	3.9	3.1	1.8	1.1	0.9	0.6	0.5
MM_1	39.2	44.5	40.7	36.6	33.4	28.3	24.8	20.9	16.8	13.6	11.2	9.5	11.6
MM_2	33.1	31.7	34.4	35.3	35.7	36.3	36.7	35.4	34.0	32.5	29.2	25.0	26.6
MM_3	19.4	10.3	12.1	15.4	18.9	23.4	28.0	32.3	37.0	40.9	45.6	49.2	50.2
MM_4	1.5	1.3	1.8	2.7	4.2	6.3	6.6	8.1	10.3	12.0	13.0	15.4	11.0

Results and Discussion

Evolution of the Serum CK-MM Isoforms after AMI and Endomyocardial Biopsy

In the 25 AMI patients treated in a conventional manner, as in those who underwent endomyocardial biopsy, the first hours of enzyme release were characterized by a rapid change in the serum MM isoform patterns. After AMI, there was a pronounced increase of the tissue isoforms, the maximum percentages being recorded 4 h after admission (table I). MM_0 and MM_1 averaged 11.9 and 44.5% of the MM activity, respectively; they represented together more than four times the percentages of MM_3 and MM_4. The MM_{-1} isoform was even detected in 2 patients (8%). In these two subjects, the serum total CK and LD levels recorded after 20 h (CK: 2,852 and 1,263 U/l; LD: 1,504 and 2,100 U/l) were higher than the mean values obtained for the other patients of the group (CK: 1,171 ± 666 U/l; LD: 1,233 ± 518 U/l). The progressive disappearance of MM_0 and MM_1 confirmed the results obtained in other series of patients [7, 8].

Thirty minutes before biopsy was started, the MM_2 and MM_3 isoforms were predominant, but after 2 h, the cellular isoforms MM_0 and MM_1 represented more than 50% of the CK-MM activity (table II). However, the percentages of the cellular isoforms regularly decreased afterwards and, after 24 h, the proportions of the five MM isoforms were approximately the

Table II. Evolution of the serum CK-MM isoforms in 17 patients who underwent endomyocardial biopsy

MM isoforms % of total CK-MM activity	Time in relation to the beginning of the procedure				
	−0.5 h	+0.5 h	+2 h	+12 h	+24 h
MM_{-1}	–	–	–	–	–
MM_0	0.5	6.5	11.2	5.6	1.5
MM_1	28.5	37.7	44.3	37.8	26.7
MM_2	35.6	33.6	31.9	38.9	38.5
MM_3	34.9	21.8	11.6	16.2	29.3
MM_4	0.5	0.4	0.6	1.5	4.0

same as those recorded in the initial samples. Although the total CK liberation was much less important in the subjects who underwent biopsy than in the AMI patients (CK peak: 112 ± 77 vs. 1,192 ± 779 U/l, $p < 0.01$), the mean proportions of the MM isoforms were similar in the two groups during the first 24 h following enzyme release; however, MM_{-1} could not be detected at any time after biopsy.

From a practical point of view, the rapid changes observed in the MM patterns of these patients clearly indicate that what we measure in serum after the release of CK from the tissues mainly consists of a complex set of degradation products resulting from the progressive alteration of the labile cellular isoforms.

Evolution of the Ratio between the Cellular and Serum Isoforms after AMI

The cellular CK-MM is characterized by three isoforms, namely MM_{-1}, MM_0 and MM_1; however, as MM_{-1} can hardly be seen after enzyme release, this isoform was not taken into account for calculating the total amount of the cellular MM fraction in the blood. Obviously representing the last stages of the MM degradation process, MM_3 and MM_4 were considered as representative of the serum isoforms.

The $(M_0 + MM_1)/(MM_3 + MM_4)$ activity ratio calculated in the 25 AMI patients treated in a conventional manner dramatically increased during the first hours after admission and reached a maximum value of 4.8 after 4 h (fig. 2). With the transformation of the cellular isoforms, the ratio decreased, a value of 1 (equivalent proportions of cellular isoforms and degradation products) being recorded between 20 and 24 h following admission. The ratio fell below 0.5 when enzyme release had begun more than 30 h earlier.

MM Isoform Patterns in AMI Patients after Thrombolytic Therapy

Early intravenous perfusion with streptokinase is now widely used in AMI patients who reach the CCU within a short time after the attack, in order to recanalize the obstructed coronary artery and, consequently, to limit the infarcted area [22]. Successful fibrinolytic therapy is known to modify the kinetics of enzyme release, because the increased washing-out from the ischaemic area leads to an earlier appearance of the tissue markers in the patient's plasma [23, 24].

The mean levels of the MM_0 and MM_{-1} isoforms were calculated in the 33 patients who received thrombolytic therapy and compared with those of

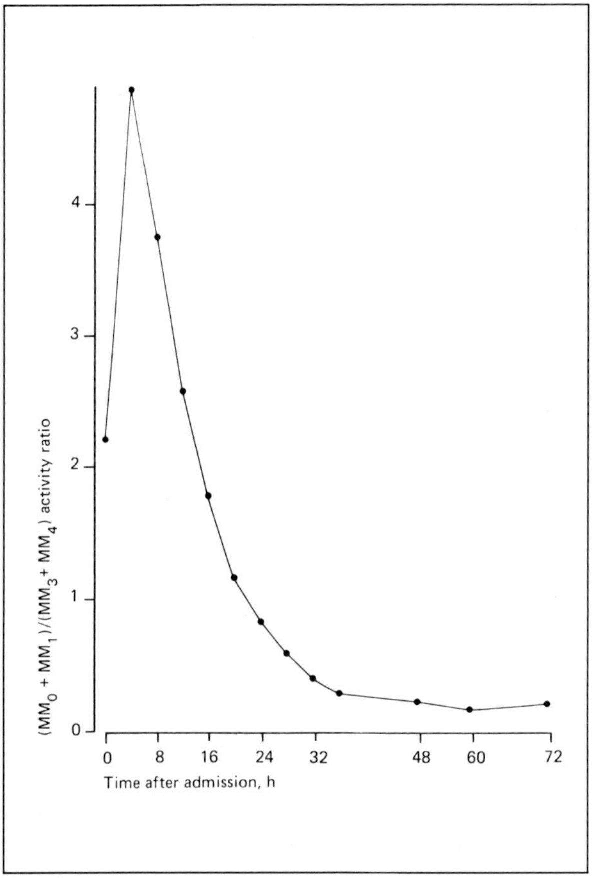

Fig. 2. Evolution of the ratio between the cellular CK-MM isoforms and their transformation products in the blood following AMI (short delays in hospitalization, intravenous heparin, n = 25).

the heparin-treated group. The only significant differences between the two groups were recorded for the MM_0 and MM_{-1} isoforms (fig. 3). MM_0 levels were significantly higher in the streptokinase-treated group 8 h after admission (13.9 ± 5.5 vs. 10.9 ± 5.1% of CK-MM, $p < 0.01$); for MM_1, the significant differences were recorded between the 4th and the 12th h; this isoform was identified in 10 patients (30%) of the streptokinase-treated group.

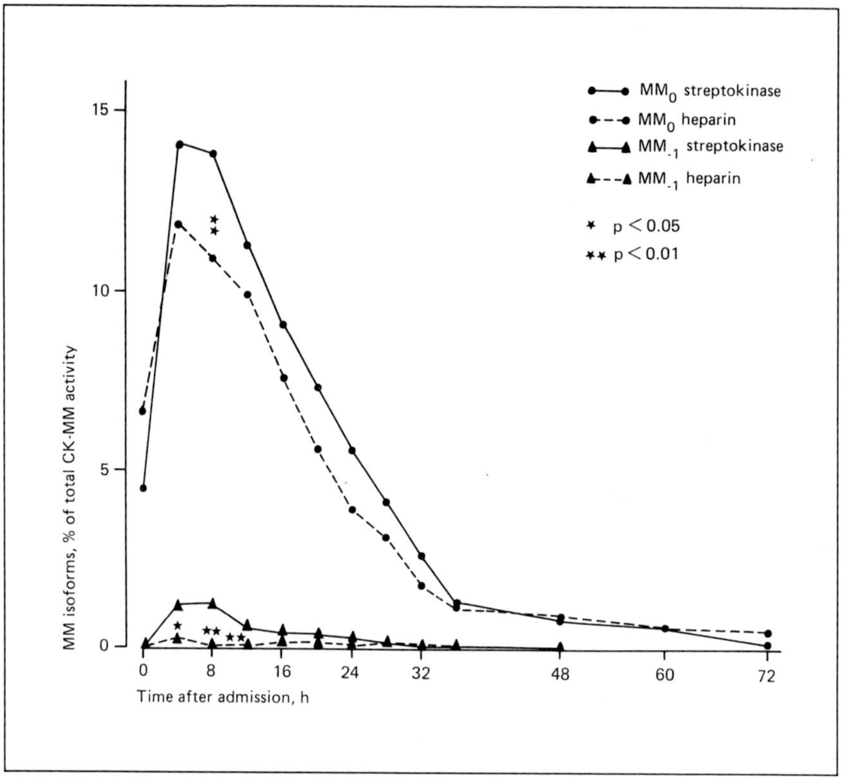

Fig. 3. Comparison of the proportions of the serum MM_{-1} and MM_0 isoforms following AMI in the streptokinase-treated group (n = 33) and in the patients who received heparin (n = 25).

Conclusion

The cellular isoforms MM_0 and MM_1 are regularly found in the blood after enzyme release from the tissues; they reach together 50% of the total MM activity 4 h after AMI, and 2 h after endomyocardial biopsy. An important finding of this study is that the minor form MM_{-1} is not only a tissue form of CK, but can be found in the patients' sera. Its occurrence, however, seems to be associated with special conditions, i.e. intense CK release after AMI, or increased washing-out in the AMI patients treated with streptokinase. Our data also confirm that the cellular forms are rapidly turned into the modified isoforms MM_2, MM_3 and MM_4, which accumulate in the blood.

As the CK-MM patterns are progressively changing in the course of AMI, the proportion of the different MM isoforms indicates the time elapsed between the beginning of the CK release and blood sampling. The $(MM_0 + MM_1)/(MM_3 + MM_4)$ ratio allows the approximate time of the onset of the symptoms to be determined, and this is particularly useful in patients who might be considered for early interventions such as coronary thrombolysis when the duration of myocardial ischaemia is in doubt.

Another application of the CK-MM heterogeneity is the development of new tissue markers, allowing a more precise delineation of the enzyme release. We have demonstrated that MM_2, MM_3 and MM_4 are not 'direct' tissue markers and MM_1 itself partly arises from MM_0 [8]; moreover, this study shows that MM_{-1} is extremely labile and can hardly be seen in the AMI patient treated in a conventional manner. Among the six MM serum isoforms, MM_0 is therefore the most appropriate for assessing the enzyme release from the tissue. Its persistence in the course of AMI indicates a continuous CK release from the myocardium, suggesting the extension of the necrotic area. Also its reappearance is an early indication of new enzyme release from the heart and accordingly of early recurrent injury.

This study also showed that the patients who received intravenous streptokinase after AMI demonstrated increased proportions of the cellular forms in the first hours following the treatment. The presence of high levels of MM_{-1} and MM_0 may also be of interest for assessing successful myocardium reperfusion after thrombolytic therapy.

The clinical applications of the heterogeneity of CK are not restricted to the management of the patients with AMI. It has also recently been demonstrated that the CK-MM isoforms were an index of disease activity in polymyositis, a MM_3/MM_1 ratio less than 1 reflecting improving or stable conditions, a MM_3/MM_1 ratio greater than 1 reflecting a deteriorating course of the disease [25]. Finally, a better comprehension of the degradation process of the MM isoforms after their release from the tissues [26, 27] provides a unique opportunity for studying the influence of the modification of the chemical structure of a molecule on the rate of its plasma clearance. Further elucidation of the mechanisms involved with enzyme removal may also be useful for the improvement of treatment of disorders associated with enzyme deficiencies.

References

1 Dawson, D.M.; Eppenberger, H.M.; Kaplan, N.O.: Creatine kinase. Evidence for a dimeric structure. Biochem. biophys. Res. Commun. *21:* 346–353 (1975).

2 Ljungdahl, L.; Gerhardt, W.: Creatine kinase isoenzyme variants in human serum. Clin. Chem. *24:* 832–834 (1978).

3 Lang, H.; Würzburg, U.: Creatine kinase, an enzyme of many forms. Clin. Chem. *28:* 1439–1447 (1982).

4 Smith, A.F.: Separation of tissue and serum creatine kinase isoenzymes on polyacrylamide gel slabs. Clinica chim. Acta *39:* 351–359 (1972).

5 Wevers, R.A.; Olthuis, H.P.; Niel, J.C.C. van; Wilgenburg, M.G.M. van; Soons, J.B.J.: A study on the dimeric structure of creatine kinase (EC 2.7.3.2). Clinica chim. Acta *78:* 271–276 (1977).

6 Wevers, R.A.; Delsing, M.; Klein Gebbink, J.A.B.; Soons, J.B.J.: Post-synthetic changes in creatine kinase isozymes (EC 2.7.3.2). Clinica chim. Acta *86:* 323–327 (1978).

7 Chapelle, J.P.; Heusghem, C.: Further heterogeneity demonstrated for serum creatine kinase isoenzyme MM. Clin. Chem. *26:* 457–462 (1980).

8 Chapelle, J.P.: Serum creatine kinase MM sub-band determination by isoelectric focusing. A potential method for the monitoring of myocardial infarction. Clinica chim. Acta *137:* 273–281 (1984).

9 Heinbokel, N.; Strivastava, L.M.; Goedde, H.W.: Agarose gel isoelectric focusine of creatine kinase (EC 2.7.3.2) isoenzymes from different human tissue extracts. Clinica chim. Acta *122:* 103–107 (1982).

10 Guslits, B.G.; Jacobs, H.K.: Investigation of the heterogeneity of the MM isoenzyme of creatine kinase. Clinica chim. Acta *130:* 55–69 (1983).

11 Yasmineh, W.G.; Yamada, M.K.; Cohn, J.N.: Postsynthetic variants of creatine kinase MM. J. Lab. clin. Med. *98:* 109–118 (1981).

12 Falter, H.; Michelutti, L.; Mazzuchin, A.; Whiston, W.: Studies on the sub-banding of creatine kinase MM and the 'CK conversion factor'. Clin. Biochem. *14:* 3–7 (1981).

13 Morelli, R.L.; Carlson, C.J.; Emilson, B.; Abendschein, D.R.; Rapaport, E.: Serum creatine kinase MM isoenzyme sub-bands after acute myocardial infarction in man. Circulation *67:* 1283–1289 (1983).

14 Chapelle, J.P.; Bertrand, A.; Heusghem, C.: The protection of creatine kinase MM sub-bands by EDTA during storage. Clinica chim. Acta *115:* 255–262 (1981).

15 Falter, H.; Michelutti, L.: An assay for creatine kinase conversion factor. Clin. Biochem. *15:* 200–205 (1982).

16 George, S.; Ishikawa, Y.; Perryman, M.B.; Roberts, R.: Purification and characterization of naturally occurring and in vitro induced multiple forms of MM creatine kinase. J. biol. Chem. *259:* 2667–2674 (1984).

17 Edwards, R.J.; Watts, D.C.: Human 'creatine kinase conversion factor' identified as a carboxypeptidase. Biochem. J. *221:* 465–470 (1984).

18 Perryman, M.B.; Knell, J.D.; Roberts, R.: Carboxypeptidase-catalyzed hydrolysis of C-terminal lysine. Mechanism for in vivo production of multiple forms of creatine kinase in plasma. Clin. Chem. *30:* 662–664 (1984).

19 Piérard, L.; El Allaf, D.; D'Orio, V.; Demoulin, J.C.; Carlier, J.: Two-dimensional echocardiographic guiding of endomyocardial biopsy. Chest *85:* 759–762 (1984).

20 Rosalki S.B.: An improved procedure for serum creatine phosphokinase determination. J. Lab. clin. Med. *69:* 696–705 (1967).
21 Wroblewski, F.; La Due, J.S.: Lactate dehydrogenase activity in blood. Proc. Soc. exp. Biol. Med. *90:* 210–213 (1955).
22 Schwarz, F.; Faure, A.; Katus, H.; Olshausen, K. von; Hofmann, M.; Schuler, G.; Manthey, J.; Kubler, W.: Intracoronary thrombolysis in acute myocardial infarction: An attempt to quantitate its effect by comparison of enzymatic estimate of myocardial necrosis with left ventricular ejection fraction. Am. J. Cardiol. *51:* 1573–1578 (1983).
23 Golf, S.W.; Temme, H.; Kempf, K.D.; Bleyl, H.; Brüstle, A.; Bödeker, R.; Heinrich, D.: Systemic short-term fibrinolysis with high dose streptokinase in acute myocardial infarction: time course of biochemical parameters. J. clin. Chem. clin. Biochem. *22:* 723–729 (1984).
24 Kwong, T.C.; Fitzpatrick, P.G.; Rothbard, R.L.: Activities of some enzymes in serum after therapy with intracoronary streptokinase in acute myocardial infarction. Clin. Chem. *30:* 731–734 (1984).
25 Annesley, T.M.; Stongwater, S.L.; Schnitzer, T.J.: MM subisoenzymes of creatine kinase as an index of disease activity in polymyositis. Clin. Chem. *31:* 402–406 (1985).
26 Abendschein, D.R.; Morelli, R.M.; Carlson, C.J.; Emilson, B.; Rapaport, E.: Creatine kinase MM isoenzyme subforms in myocardium, cardiac lymph and blood after coronary artery occlusion in dogs. Cardiovasc. Res. *18:* 690–696 (1984).
27 Hashimoto, H.; Grace, A.M.; Billadello, J.J.; Gross, R.W.; Strauss, A.W.; Sobel, B.E.: Nondenaturing quantification of subforms of canine MM creatine kinase isoenzymes (isoforms) and their interconversion. J. Lab. clin. Med. *103:* 470–484 (1984).

J.-P. Chapelle, MD, Laboratoire de Chimie Médicale, Université de Liège,
Rue des Bonnes-Villes 1, B–4020 Liège (Belgium)

Adv. clin. Enzymol., vol. 3, pp. 116–125 (Karger, Basel 1986)

Multiple Forms as an Index of the Age of an Enzyme

A Review of Postsynthetic Modification in Serum by the Modifying Protein

R.A. Wevers[a], *J.B.J. Soons*[b]

[a]Department of Clinical Chemistry, Institute of Neurology, Radboud Hospital, University of Nijmegen, The Netherlands;
[b]State University Utrecht, The Netherlands

Protein Multiplicity

It is generally known that many enzymes and proteins occur in various forms within one species. In many cases the explanation for the existence of this enzyme multiplicity can be found in the genetic material. The term 'isoenzymes' applies in these cases [1].

Another possibility is that a postsynthetic modification takes place in the protein, by which the original protein as it was coded for in the genes is later processed or modified. The term 'multiple forms' has to be used in these instances [1]. There may be various reasons why the protein structure of a specific enzyme is modified. Sometimes the enzyme is produced in an inactive form and the modification is necessary to obtain the desired catalytic activity. Other enzymes are processed in order to reach a specific localization within the body. They are targetted to a certain organ or organelle by the modification process. In fact there are many more reasons for postsynthetic modifications that will not be discussed here.

The site in the body where the modification process takes place varies considerably. In some instances the modification is confined to a certain well-defined intracellular location [2, 3]. Other enzymes are exclusively modified in the blood. This report deals with examples from this category. It describes a normal serum protein, capable of bringing about postsynthetic alterations in several human enzymes and isoenzymes. Creatine kinase (CK) and the glycolytic enzyme enolase are modified by this protein, for which the descriptive name 'modifying protein' is used here.

Examples of the Action of the Modifying Protein

Creatine Kinase

Creatine kinase is a dimeric enzyme. Three cytosolic isoenzymes are known to occur. They have been named CK MM, MB and BB. Because of their tissue specificity the isoenzymes are widely used for diagnostic purposes in clinical chemistry. The most important application is the use of CK MB as a marker for cardiac tissue damage. In the 1970s it was demonstrated that the CK M-chain undergoes a postsynthetic modification. Three CK MM forms and two CK MB forms were found to occur in the blood [4–8].

Extracts of human muscular tissue proved to contain only one of the three MM forms and maximally one of the two MB forms [6]. This finding already suggested that the modification process takes place outside the muscular tissue. Finally, it could be demonstrated to happen in the blood [6, 7, 9–11]. The various forms of CK MM and CK MB are multiple forms. Obviously, the modification that takes place here is a postsynthetic process.

Table I gives the nomenclature that is in use for these multiple forms and explains the subunit composition [4–6]. Generally accepted rules for enzyme nomenclature have been applied. Accordingly, the CK MM form

Table I. Nomenclature and physical properties of the various CK multiple forms

Nomenclature	Chain composition	Isoelectric point	Activation energy	K_m (25 OC)	
				ADP	CrP
CK MM 3	original tissue form	6.89	49.1	0.22	3.3
CK MM 2	1 M chain modified	6.51	59.2	0.19	2.6
CK MM 1	2 M chains modified	6.23	66.8	0.17	2.1
CK MB 2	original tissue form	n.d.	n.d.	n.d.	n.d.
CK MB 1	1 M chain modified	n.d.	n.d.	n.d.	n.d.

n.d. = Not determined.

with the highest anodal mobility upon electrophoresis should be called CK MM1. Unfortunately, the original tissue forms (CK MM3 and CK MB2) have the highest cathodal mobility upon electrophoresis. This has led some authors to use another system of nomenclature [9] and in doing so they have assigned the lowest numbers to the original tissue forms. Although it seems not illogical to call the original tissue form of the enzyme MM1, this is rather confusing in view of the international rules and should therefore be avoided.

Furthermore, the table shows that the modification process also alters some of the physicochemical properties of the enzyme. For CK MM it could be demonstrated that the isoelectric point [5, 9], the apparent activation energy [12] and the Michaelis-Menten constants [12] of the enzyme are affected by the alteration in the molecule. No alterations were found in the behaviour of the various MM forms against a polyclonal antiserum [7]. All three CK MM forms mentioned in table I occur in normal human serum. This implies that the alterations in the two M chains of the molecule do not take place simultaneously. The MM form with one original and one modified chain (CK MM2) is stable. In fact, it proved to be possible to produce CK MM2 in vitro from purified CK MM3 and CK MM1 by means of hybridization experiments [5]. Final proof for the chain composition described in table I came from similar hybridization studies [5]. The mechanism of the CK modification was unknown for a long time. In 1981, it was recognized for the first time that a normal (Ca-dependent) human serum protein was responsible for the modification [7, 10, 11]. Several names were used for this, as yet unknown, protein: CK conversion factor [10], modifying protein [13] and, finally, carboxypeptidase K [14]. The latter name was given after unravelling the molecular basis of the CK modification by Edwards and Watts [14] in 1984. The carboxy-terminal lysine amino acid of the CK M chain was found to be removed by the modifying protein, suggesting it to be a carboxypeptidase [14, 15].

Some authors have described the existence of more than three CK MM multiple forms. Many subspecies of CK MM have been found in muscular tissue [8, 16–18]. Guslits and Jacobs [16] even found 21 MM subspecies, the highest number reported in the literature so far.

Varying numbers of MM multiple forms have also been described for serum. Wevers et al. [5] described a CK X band in human serum using column isoelectric focusing techniques, bringing the total of CK MM subbands to four. This is in agreement with the study by Yasmineh et al. [19]. Chapelle and Heusghem [9] on the other hand described a total of five MM multiple forms in human serum and found that two of these could also be demonstrated in muscular tissue extracts. It is obvious that the traditional dimeric structure and even the new extended dimeric structure outlined above do not satisfactorily explain the existence of more than three CK MM and two CK MB forms. All authors agree, however, that there is a process of postsynthetic modification taking place in the circulation which causes the original tissue form(s) of the enzyme to be modified.

Enolase

The glycolytic enzyme enolase has a dimeric structure. Three different subunits, α, β and γ, are known and five homo- and heterodimeric combinations are known to occur in vivo. The homodimeric forms αα, ββ and γγ are widely distributed in the body and have an obvious tissue specificity. ββ is the predominant isoenzyme form in muscular tissue, while αα and γγ both occur in important amounts in brain tissue. The enzyme became of renewed interest after the discovery of separate locations for αα and γγ in brain tissue. The former turned out to be the major isoenzyme in glia cells while the latter was present in the neurons and in neuronal tissue [20]. Several attempts have been made to use these isoenzymes as diagnostic markers for damage affecting specific cell types.

Marangos et al. [21] were the first to obtain experimental evidence for the existence of multiple forms of αα enolase. Subsequently, it could be demonstrated that the isoenzyme actually undergoes a postsynthetic modification and that this modification is brought about by the same modifying protein that alters the structure of creatine kinase [13]. Recently, we have found that the ββ form of enolase can also be modified by the purified modifying protein. For both isoenzymes the mechanism of the modification process is identical to the mechanism of CK MM modification. For both αα and ββ enolase, three multiple forms can be discerned, indicating that forms with one original and one modified chain exist. All multiple forms of this enzyme have actually been found in patients' sera. In brain and muscular tissue extracts the unmodified forms of the isoenzymes (αα 3 and ββ 3) are the predominant forms. Minor quantities of αα 2 sometimes found in brain tissue extracts [13] can probably be explained by serum contamination during the extraction procedure.

Again, as for CK, the physical properties of αα and ββ enolase are altered by the modification [for details see ref. 9]. Table II compares the direction of changes in physical parameters of CK and the enolases. A consistent decrease of pI is observed upon modification. The activation energy,

Table II. Changes in physicochemical parameters brought about by the modification process

	Changes brought about by modification			
	pI	K_m	activation energy	antibody affinity
CK MM	↓	↓	↑	unchanged
Enolase αα	↓	minor decrease	↓	n.d.
Enolase ββ	↓	n.d.	↓	n.d.

n.d. = Not determined.

however, increases in case of CK and decreases for the enolases by the alteration in the enzyme. The fact that the physical properties of the enzyme under modification often change can be used as an easy screening tool in finding other enzymes or proteins that are modified.

The Modifying Protein

In 1978 it was found that the factor responsible for the modification in the CK-M chain is thermolabile [6] and in 1981 it was described as a Ca-dependent serum protein [7, 10]. It has an isoelectric point of 4.35 [22]. Estimations of its molecular mass vary between 185 and 190 kdaltons [10, 22]. Falter et al. [11] described an assay for the determination of the modifying protein. The assay seems to have the disadvantage of not being completely specific. The modifying protein was at first partially purified [10, 13, 14], and later highly purified preparations became available [22]. As yet, no publications have appeared about antibody production against the modifying protein. Two independent groups have confirmed the molecular mechanism of the modifying protein [14, 15]. In the case of CK MM it was found to act as a carboxypeptidase, splitting off the carboxy-terminal lysine residue of the M chain [14, 15]. In 1967, Yue et al. [23] had already shown that lysine is the carboxy-terminal amino acid of CK in rabbit skeletal muscle. Perryman et al. [15] also found lysine at the carboxy-terminal end of the human CK M chain. Like the modifying protein, both the known carboxypeptidases A and B have been found to be able to split off the

Table III. Modifying activity in rat tissues

Tissue	Present?	Tissue	Present?
Serum	+	Brain	−
Muscle	−	Pancreas	+
Heart	−	Bone marrow	−
Liver	−	Thymus	−
Spleen	−	Adrenal gland	−
Kidney	−	Hypophysis	−

C-terminal lysine [14, 15]. However, based on obvious differences in molecular mass between the modifying protein and the known carboxypeptidases, it is assumed that they are dissimilar proteins (molecular mass: (1) CpA and CpB: 30–40 kdaltons (2) modifying protein 185–190 kdaltons [10, 22]). Perryman et al. [15] report that carboxypeptidase N is also capable of removing the C-terminal lysine from the human CK M chain, and suggest that the modifying protein could be carboxypeptidase N. Data about the estimated molecular mass of CpN seem to contradict this (molecular mass CpN: 280 kdaltons [24]). The modifying protein could indeed be a new carboxypeptidase. However, it has not yet been shown that its action on any other protein can also be described as a carboxypeptidase action. Moreover, it has not yet been excluded that the modifying protein is a complexed form of the known carboxypeptidases. We suggest use of the name modifying protein until more solid evidence has been found to call it either carboxypeptidase K as proposed by Edwards and Watts [14] or carboxypeptidase N as suggested by Perryman et al. [15].

The tissue of origin of the modifying protein has not yet been reported. Table III shows the results of an experiment through which we tried to answer this question in the rat model (inbred Wistar strain). After incubating tissue extracts from the rat with either rat or human enolases, we subsequently checked the incubation mixture electrophoretically for modifications. We were able to detect the modifying protein in serum of the rat, thereby extending its occurrence to other species. The only other organ that we have found to contain modifiying activity in the rat was pancreatic tissue. This proved to be modifying activity caused by the carboxypeptidases A and B as judged by its molecular mass (measured molecular mass: 37.5 kdaltons). Although we have tested the most probable tissues of origin of the modifying protein we have not succeeded in finding the tissue where

Table IV. Comparison of proteins modified by the modifying protein

	Number of chains	Tissue of origin	Molecular mass kdaltons	pI
CK MM	2	muscle heart	82	6.9
CK MB	2	heart	82	5.2
Enolase αα	2	liver brain	90	7.0
Enolase ββ	2	muscle heart	90	7.0–8.0

it is produced. Several possible explanations can be thought of. The protein could be secreted by a cell type that has not been investigated or it could be produced in an inactive form in the tissue of origin. The possibility that the modifying activity has been lost in the extraction procedure was excluded by adding known amounts of the purified protein to the various tissue extracts and studying its recovery. A third possibility would be that the modifying protein actually is built up by complex formation from the known carboxypeptidases or actually is carboxypeptidase N [15]. Antibody studies seem to be the only solution in finding the answers to the questions raised here.

Perhaps the most important unresolved question is the in vivo function of these enzyme modifications. While we only knew that the modifying protein alters the structure of CK it was tempting to think in terms of CK as an 'innocent bystander'. This becomes less likely when more proteins prove to be modified by the action of this protein. The list of actual biological substrates of the modifying protein is by no means complete yet. The enolase isoenzymes αα and ββ have taught us that the action of the modifying protein is not confined to only one enzyme. Still, there are obvious similarities between CK MM and the enolase isoenzymes (table IV), showing that we still cannot completely rule out the 'innocent bystander' theory. It is important to keep on searching for other proteins serving as a substrate for the modifying protein. In this respect the suggestion that certain haemoglobin forms could be candidates for modificatory changes by the modifying protein is very important [14]. Could it be that the common denominator among the substrates is that they actually do not belong in the circulation

Table V. CK apparent activation energy (in kJ/mol) in subsequent serum samples from eight patients after myocardial infarction

	Time after infarction					
	0–12 h	13–24 h	25–36 h	37–48 h	49–60 h	61–72 h
CK activation energy	55.1	58.8	61.2	63.0	63.8	66.8
SD	1.8	1.3	2.3	1.4	1.7	1.0
N	12	5	10	4	6	7

and that they have to be removed? As yet there are no indications that the modification serves to inactivate the proteins involved. The modifying protein does not directly inactivate the potentially harmful enzymatic activity of the enzymes. It could, however, well be that it indirectly prepares the enzyme for a faster clearance from the circulation. Is the process of modification perhaps a starting signal for a faster clearance of certain proteins from the blood?

A Practical Application

As the modifying protein occurs in every human serum tested so far, it must be considered a normal human serum protein. The earliest papers [4, 9, 10] already showed that the CK MM pattern after a myocardial infarction changes gradually during a period of roughly 60 h [4, 9, 10]. This implied that it would be possible to estimate how long the enzyme had been in the circulation. Such estimates can only be valid if the interindividual variation in the quantity of modifying protein in serum is not too large. Recently [12, 25], this has been confirmed once more using the activation-energy of CK as an index of the age of the enzyme in the blood (table V). Of course, similar results could have been obtained using electrophoretic procedures. From a practical point of view the measurement of the activation energy is easier to perform. It is very interesting that such measurements could theoretically open new ways of detecting myocardial reinfarction or myocardial infarction in a post-operative state.

For clinical chemistry in general, the results in this patient-model of myocardial infarction look very promising. As we may expect to find more and more enzymes and proteins that serve as a substrate for the modifying protein it is very probable that the example presented here is the first of many more in which we can begin to estimate how long certain proteins have been in the circulation. These estimates can be made from measurements in only one serum sample from the patient.

References

1 IUPAC-IUB: Nomenclature of multiple forms of enzymes. Recommendations 1976. Clin. Chem. *23:* 2163–2165 (1977).

2 Rothstein, M.: Recent developments in the age related alterations of enzymes. A review. Mech. Age. Dev. *6:* 241–257 (1977).

3 Womack, J.E.: Post-translational modification of enzymes: processing genes; in Isozymes: current topics in biological and medical research, vol. 7. Molecular structure and regulation, pp. 175–186 (Liss, New York 1983).

4 Wevers, R.A.; Olthuis, H.P.; Niel, J.C.C. van; Wilgenburg, M.G.M. van; Soons, J.B.J.: A study on the dimeric structure of creatine kinase. Clinica chim. Acta *75:* 377–385 (1977).

5 Wevers, R.A.; Wolters, R.J.; Soons, J.B.J.: Isoelectric focusing and hybridization experiments on creatine kinase. Clinica chim. Acta *78:* 271–276 (1977).

6 Wevers, R.A.; Delsing, M.; Klein Gebbink, J.A.G.; Soons, J.B.J.: Post synthetic changes in creatine kinase isozymes. Clinica chim. Acta *86:* 323–327 (1978).

7 Wevers, R.A.: Creatine kinase. Isozymes and multiple forms; thesis, State University Utrecht (1981).

8 Smith, A.F.: Separation of tissue and serum creatine kinase isoenzymes on polyacrylamide gel slabs. Clinica chim. Acta *39:* 351–359 (1972).

9 Chapelle, J.P.; Heusghem, C.: Further heterogeneity demonstrated for serum creatine kinase isoenzyme MM. Clin. Chem. *26:* 457–462 (1980).

10 Falter, H.; Michelutti, L.; Mazzuchin, A.; Whiston, W.: Studies on the sub-banding of creatine kinase MM and the 'CK conversion factor'. Clin. Biochem. *14:* 3–7 (1981).

11 Falter, H.; Michelutti, L.; Mazzuchin, A.: An assay for the creatine kinase conversion factor. Clin. Biochem. *15:* 200–205 (1982).

12 Wevers, R.A.; Hagelauer, U.; Stein, W.; Bohner, J.; Faust, U.; Landeghem, A.A.J. van; Soons, J.B.J.: Indices for the age of the creatine kinase M-chain in the blood. Clinica chim. Acta *148:* 197–210 (1985).

13 Wevers, R.A.; Boegheim, J.P.J.; Hommes, O.R.; Landeghem, A.A.J. van; Mul-Steinbusch, M.W.F.J.; Stappen, J.W.J. van der; Soons, J.B.J.: A study on post-synthetic modifications in alfa-alfa enolase (EC 4.2.1.11.) brought about by a human serum protein. Clinica chim. Acta *139:* 127–135 (1984).

14 Edwards, R.J.; Watts, D.C.: Human 'creatine kinase conversion factor' identified as a carboxypeptidase. Biochem. J. *221:* 465–470 (1984).

15 Perryman, M.B.; Knell, J.D.; Roberts, R.: Molecular mechanism for the production of multiple forms of MM creatine kinase. Experientia *40:* 1275–1277 (1984).

16 Guslits, B.G.; Jacobs, H.K.: Investigation of the heterogeneity of the MM isozyme of creatine kinase. Clinica chim. Acta *130:* 55–69 (1983).
17 Traugott, C.; Massaro, E.J.: The molecular basis of heterogeneity of the MM isozyme of rabbit muscle creatine phosphokinase. Biochim. biophys. Acta *295:* 549–554 (1973).
18 Rosenberg, B.; Eppenberger, H.M.; Perriard, J.C.: Occurrence of heterogeneous forms of the subunits of creatine kinase in various muscle and non-muscle tissues and their behaviour during myogenesis. Eur. J. Biochem. *116:* 87–92 (1981).
19 Yasmineh, W.G.; Yamada, M.K.; Cohn, J.N.: Post-synthetic variants of creatine kinase MM. J. Lab. clin. Med. *98:* 109–118 (1981).
20 Schmechel, D.; Marangos, P.J.; Zis, A.P.; Brightman, M.: Brain enolases as specific markers of neuronal and glial cells. Science *199:* 313–315 (1978).
21 Marangos, P.J.; Parma, A.M.; Goodwin, F.K.: Functional properties of neuronal and glial isoenzymes of brain enolase. J. Neurochem. *31:* 727–732 (1978).
22 Van Landeghem, A.A.J.; Soons, J.B.J.; Wevers, R.A.; Mul-Steinbusch, M.W.F.J.; Antonisse-Zijda, T.: Purification and determination of the modifying protein responsible for the post-synthetic modification of creatine kinase and enolase. Clinica chim. Acta *153:* 217–224 (1985).
23 Yue, R.H.; Palmieri, R.H.; Olson, O.E.; Kuby, S.A.: Studies on adenosine triphosphate transphosphorylases V. Biochemistry *6:* 3204–3227 (1967).
24 Plummer, T.H.; Hurwitz, M.Y.: Human plasma carboxypeptidase N. J. biol. Chem. *253:* 3907–3912 (1978).
25 Hagelauer, U.; Faust, U.: The catalytic activity and activation energy of creatine kinase isoenzymes. J. clin. Chem. clin. Biochem. *1982:* 633–638.

R.A. Wevers, MD, Department of Clinical Chemistry, Radboud Hospital,
Institute of Neurology, University of Nijmegen, Reinier Postlaan 4, PO Box 9101,
NL–6500 HB Nijmegen (The Netherlands)

Adv. clin. Enzymol., vol. 3, pp. 126–132 (Karger, Basel 1986)

Current Methods for the Determination of Creatine Kinase-MB Catalytic Activity

D. Hollez, J. Spincemaille, V. Blaton

Department of Clinical Chemistry, AZ Sint-Jan, Brugge, Belgium; Department of Clinical Chemistry, Institute Pharmacological Science, KU-Leuven, Belgium

Creatine kinase (CK), although present in low concentrations in most of the body tissues, is found in the highest concentrations in the intracellular fluid of the brain, skeletal muscle and cardiac muscle. Creatine kinase isoenzymes result from the non-covalent association of the subunits into all three possible combinations. The BB isoenzyme, referred to as CK-1, has the highest electrophoretic mobility and is found primarily in brain. The isoenzyme CK-MB (CK-2) has an intermediate electrophoretic mobility and accounts for approximately 40% of the CK present in heart extracts, the other significant source of CK-MB being skeletal muscle, in which it accounts for at least 4% of the total CK activity. Serum CK-MB concentrations are now used to determine cardiac damage, such as myocardial infarction [1, 2]. Moreover, the concentration of the CK-MB subfraction has been used to assess the extent of the myocardial infarction and subsequent reinfarction.

The early assays used for the detection of CK isoenzymes were dependent upon physical separation techniques, in particular electrophoresis or ion-exchange column chromatography followed by assessment of CK activity in the isolated fractions [3, 4]. Radioimmunoassays (RIA) and immunoinhibition assays have been developed for the measurement of CK isoenzymes [5, 6]. Column chromatography is quantitative and sensitive, but does not lend itself to the assay of large number samples. Several immunochemical techniques have been reported. At present, there are three types of such assays: immunoinhibition of the M subunit, RIA of the B subunit and RIA of the CK-MB isoenzyme.

In the immunoinhibition method, antibodies of the MM isoenzyme inhibit M subunit activity of both the CK-MM and CK-MB isoenzymes. In the RIAs, labelled BB isoenzyme competes with the MB and (or) BB isoenzymes in serum for binding sites on a limited number of BB antibodies. The immunoinhibition and RIA methods therefore cannot differentiate the MB isoenzyme in the presence of the BB isoenzyme and (or) any variant CK isoenzymes containing BB. They are not MB specific; they simply give a measurement of B subunit activity. An immunochemical, MB-specific procedure is the radiometric test reported by Usetagui-Gomez et al. [7]. Electrophoresis has been widely used, but the occurrence of immunoglobulin-bound CK-MM and CK-BB and mitochondrial CK variants, as well as fluorescent artefacts, which migrate near the CK-MB region, necessitate careful interpretation to avoid false-positive results. The ion-exchange procedures likewise demonstrate frequent false-positive results. The RIA assays demonstrate greater sensitivity and immunoinhibition offers more technical simplicity; both, however, are dependent upon the production of antisera which recognize either the M or the B subunit of the CK-MB molecule. These techniques are subject to false-positives, including those produced by elevated CK-BB, or free B subunit.

A novel immunochemical assay for a specific enzymic determination of the myocardial isoenzyme of CK-MB involves determination of the B subunit activity inhibited by specific antibodies to the M subunit. Interfering activities from CK-BB isoenzyme, atypical forms of creatine kinase, and adenylate kinase are eliminated by using a blank tube in which all the M subunit-containing isoenzymes have been removed by a specific immunoprecipitation step. The assay is convenient, linear, and reproducible, and results compare well with those obtained by agarose electrophoresis [6].

Immunoenzymetric assays, now available, use two monoclonal antibodies to measure the intact CK-MB molecule with high sensitivity and specificity, which is not affected by increased concentrations of CK-MM or CK-BB. One monoclonal antibody is directed towards the M subunit, producing an assay which is specific for CK-MB.

We evaluated and compared in a first series of experiments three main current methods for the determination of the creatine kinase MB activity: the electrophoretic system (Beckman Inst.), the ion-exchange MB-Zyme (Baker) and the ion-exchange on aca (Dupont). Patients were carefully selected in order to obtain information on the specificity, sensitivity and efficiency of these methods in the diagnosis of acute myocardial infarction classified according to the WHO diagnostic criteria and confirmed by our department of cardiology.

A more direct assay for CK-MB is based upon the use of two monoclonal antibodies. In our present work we compared the CK-MB values determined by immunoinhibition and by an immunoenzymatic assay on patients with a positive CK-MB electrophoretic pattern. Special attention was given to patients with aberrant creatine kinase isoenzymes. Our purpose was to evaluate the analytical performances of the CK-MB immunoenzymatic assay in terms of precision and linearity.

Comparative Data of CK Isoenzymes Obtained by Physical and Immunochemical Methods

CK-MB Isoenzyme Assay Dependent upon Physical Separation Techniques

Three current methods for the determination of the CK-MB activity were first evaluated: electrophoresis (Beckman Inst.), ion-exchange MB-Zyme (Baker) and ion-exchange on aca (Dupont). The limitations of each method were investigated by comparing the data with those obtained from the combination of methods and the analytical precision data were tested on the two standards with high and low CK-MB values. We collected 80 samples from 80 selected in-patients. Fifty were confirmed myocardial infarctions, 3 were known to have a persistent macro CK type I and 27 had no proven myocardial infarction, from which a significant number (n = 15) came from the vascular surgery unit. From acute myocardial infarction patients, blood samples were collected between 12 and 24 h after onset. Whole blood was centrifuged at 1,200 *g* and CK was determined immediately, or later on frozen samples kept at –25 °C according to a modified method of Rosalki [8]. The quantitative isoenzyme electrophoresis was carried out on Paragon™ CK (Beckman Inst., Inc., Brea, Calif.) and scanned by increase of NADPH fluorescence on a Microzone Densitometer (Beckman Inst. R-112, Inc., Fullerton, Calif.). The ion-exchange method MB-Zyme (Baker Chemicals B.V., Deventer, Holland) was used strictly according to the procedures of the manufacturer on an UV spectrophotometer (Beckman Inst., Inc., Iwine, Calif.). Macro CKs (IgG-bound CK) were determined by a precipitation technique with protein-A-Sepharose CL-4B (Pharmacia Fine Chemicals AB, Uppsala, Sweden). To confirm these

Table I. Comparative prediction data of three methods for CK-MB in a group of 77 selected patients

Methods	Sensitivity	Specificity	Efficiency
aca Dupont	98	74	88
MB-Zyme	94	92	93
Electrophoresis	86	89	87
CK total	98	51	82

$$\text{Sensitivity} = \frac{\text{correctly classified AMIs}}{\text{total AMIs}} \times 100.$$

$$\text{Specificity} = \frac{\text{correctly classified non-AMIs}}{\text{total non-AMIs}} \times 100.$$

$$\text{Efficiency} = \frac{\text{true positives + true negatives}}{\text{total patients}} \times 100.$$

forms, activation energy was calculated after measuring the enzyme catalytic activity at 25 and 37 °C [9].

The three methods described were tested on patients' sera (n = 77) and on two pools: one from a patient group with a total CK of less than 200 U/l and the other obtained from a group with more than 200 U/l. The precision data for the three methods with respect to those pools, respectively, are for the aca Dupont 13.9% CV and 5.7%, and for the MB-Zyme method, 13.3% and 5.2%. Electrophoresis overestimated both the low and high activity pools. The ion-exchange methods showed an increased precision for the high activity measurements. The clinical usefulness of the three methods is determined by their specificity, efficiency and sensitivity as shown in table I. The efficiency of the three methods is comparable. An excellent sensitivity is noted for the aca Dupont system, while the Beckman electrophoresis showed a lower value due to the fact that very low creatine kinase-MB values could not be detected by this method. Electrophoresis is still the best method for detecting macro CKs [10].

Immunoenzymetric Assay of CK-MB

A more direct assay for the CK-MB is based on the use of two monoclonal antibodies, one against the M subunit and one against the B subunit. The present work was undertaken to compare the CK-MB values determined by immunoinhibition and monoclonal antibodies on patients with a positive CK-MB band on the electrophoretic pattern. Special attention was given to patients with aberrant CK isoenzymes; 20 patients with macro CK type I were studied. Our purpose was to evaluate the analytical performance of the Tandem-E CK-MB technique in terms of precision and linearity. In this technique, samples containing CK-MB are reacted with a plastic bead (solid phase) coated with a monoclonal antibody directed towards a unique antigenic site on the M subunit of the CK-MB molecule, and with an enzyme-labelled monoclonal antibody directed against an antigenic site on the B subunit of the CK-MB molecule. After the formation of the solid phase (CK-MB) labelled antibody sandwich, the bead is washed to remove the unbound labelled antibody and then is incubated with enzyme substrate (*p*-nitrophenyl phosphate). The amount of substrate converted is determined colorimetrically. Within-run precision of the assay was evaluated by measuring a sample with a CK 1,420 U/l and a CK-MB value of 20% of total CK on electrophoresis. The CV was 4.6%. Between-run precision was evaluated on a serum pool with a CK activity of 1,320 U/l and a CK-MB value of 15% of total CK and was analyzed over 8 days. We obtained mean levels quite similar to those calculated during the within-run precision experiment, but with higher dispersion (CV, 11%). Incubation and rotation of the beads during the assay play an important role in determining the CV%.

In the 34 control subjects, total CK averaged 64 ± 26 IU/l and the CK-MB activity using the immunoinhibition technique ranged 15.6 IU/l. The CK-MB concentration determined by the Tandem-E CK-MB assay averaged 1.32 ng/ml (range 1–7 ng/ml).

In a comparative study on specificity, three current methods for the determination of CK-MB activity were used and evaluated: the described electrophoresis method, the immunoinhibition test (Boehringer, Mannheim) and the immunoenzymatic monoclonal assay. Table II describes the data obtained on 25 patients which were positively screened on electrophoresis with a CK-MB value higher than 5%. All patients give a positive result higher than 6 ng/ml, which is the normal value obtained with the immunoenzymatic assay. Although electrophoresis correlates well with the ion exchange methods, a lower correlation was obtained with the immunoinhibition test (r = 0.7656) and with the monoclonal assay (r = 0.6725): an identical trend is seen both for low and high CK values. In table II we calculated the regression line between the CK immunoinhibition test and the immunoenzymatic assay, and we noted a small correlation (r = 0.67982)

Table II. CK-MB values in randomized serum samples assayed with the immunoinhibition method and with the immunoenzymatic monoclonal technique

	Total CK, mU/ml	Immunoinhibition (a) CK-MB, mU/ml	Monoclonal (b) CK-MB, ng/ml
n	25	25	25
Mean	543	66.3	55.6
± SD	362.65	75.61	40.48
SEM	72.53	15.12	8.10

Correlation coefficient between (a) and (b): r = 0.67982.
Regression line: Y = 0.3640 X + 31.5246.

Table III. CK-MB values assayed on serum samples with atypical CK isoforms

	EP CK-MB, %	CK, mU/ml	CK-MB, mU/ml	CK-MB, ng/ml
n	16	16	16	16
Mean	15	357	112	1.7
± SD	12.63	961.40	169.40	2.16
SEM	3.16	240.35	42.35	0.54

between both methods. Regarding the clinical data in the same group of 25 patients, we obtained no false-positives or false-negatives with the immunoenzymatic method.

During our experiments special attention was given to the evaluation of CK-MB in patients with aberrant CK isoenzymes. Sixteen patients with a macro CK type I were selected and their plasma was assayed for CK-MB by electrophoresis, immunoinhibition and immunoenzymatic assay. The results are described in table III. The macro CK type I samples are positively evaluated with the immunoinhibition test and negatively with the monoclonal assay, except for two samples. With affinity chromatography, both samples were further studied to separate the IgG-CK-BB complexes. In both cases we were unable to isolate completely the macroforms, indicating that in both samples the macro CK type I contained IgA-CK-BB besides IgG-CK-BB and also some free CK-MB present in the complex.

Our data showed advantages of using the monoclonal assay in the evaluation of serum samples for CK-MB isoenzyme. Although CK-MB concentrations (immunoenzymatic) and activities (immunoinhibition) showed a quite parallel evolution in the course of myocardial infarction, it is difficult to define a unique conversion factor between CK-MB mass and CK-MB activity.

The immunoenzymatic method provides an excellent discrimination between normal and pathological values and is promising to evaluate acute myocardial infarction or damage after cardiac surgery, by precisely determining the CK-MB released in the blood during the intervention.

References

1 Galen, R.S.; Reiffel, J.A.; Gambino, S.R.: Diagnosis of acute myocardial infarction. J. Am. med. Ass. *232:* 145–148 (1975).

2 Dillon, M.C.; Calbreath, D.F.; Dixon, A.M.; Rivin, B.E.; Roark, S.F.; Ideker, R.E.; Wagner, G.S.: Diagnostic problem in acute myocardial infarction. Archs intern. Med. *142:* 33–38 (1982).

3 Hamlin, C.; Ackerman, E.: Relative merits of two electrophoretic and two column-chromatographic kits for determining serum creatine kinase isoenzyme MB activity. Clin. Chem. *24:* 2013–2018 (1978).

4 Shell, W.E.; Kligerman, M.; Rorke, M.P.; Burnam, M.: Sensitivity and specificity of MB creatine kinase activity determined with column-chromatography. Am. J. Cardiol. *44:* 67–71 (1979).

5 Ritter, C.S.; Mumm, S.R.; Roberts, R.: Improved radio-immunoassay for creatine kinase isoenzymes in plasma. Clin. Chem. *27:* 1878–1881 (1981).

6 Wicks, R.; Usategui-Gomez, M.; Miller, M.; Warshow, M.: Immuno-chemical determination of CK-MB isoenzyme in human serum. II. An enzymic approach. Clin. Chem. *28:* 54–58 (1982).

7 Usategui-Gomez, M.; Wicks, R.W.; Farrenkopf, B.: Immuno-chemical determination of CK-MB isoenzyme in human serum. A radiometric approach. Clin. Chem. *27:* 823–827 (1981).

8 Rosalki, S.B.: An improved procedure for serum creatine phosphokinase determination. J. Lab. clin. Med. *69:* 696–705 (1967).

9 Stein, W.; Bohmer, J.; Steinhart, R.; Eggstein, M.: Macro creatine kinase. Determination and differentiation of two types by their activation energies. Clin. Chem. *28:* 19–24 (1982).

10 Spincemaille, J.; Delanghe, J.; De Buyzere, M.; Breemeersch, M.; Blaton, V.: Evaluation of three current methods for the determination of creatine kinase-MB catalytic activity. J. clin. Chem. clin. Biochem. *22:* 603–607 (1984).

V. Blaton, Dr. sc., Department of Clinical Chemistry, A.Z. Sint-Jan, Ruddershove 10, B–8000 Brugge (Belgium)

Adv. clin. Enzymol., vol. 3, pp. 133–138 (Karger, Basel 1986)

Specific Immunoassay of Alpha-Amylase Isoenzymes in Human Serum

K. Wulff, M. Gerber

Boehringer Mannheim GmbH, Research Center Tutzing, Tutzing, FRG

There are two isoenzymes of α-amylase present in human serum, the salivary and the pancreatic α-amylases. They are glycoproteins and show their main differences in the glycosylated part of the enzyme molecules [1]. Recent studies on the cDNA sequences of the respective mRNAs show only about 7% difference between the amino acid sequences of the two isoenzymes [2].

The activities of lipase and pancreatic α-amylase in serum are the most important parameters for the diagnosis of pancreatitis. The analysis of pancreatic α-amylase can be performed by electrophoresis [3], electrofocusing [4], or by inhibiting the salivary enzyme by wheat germ proteins [5, 6].

Since these methods are either time-consuming or show a low precision, the total serum amylase activity is assayed usually in the routine laboratory.

Recently, we described a monoclonal antibody which specifically binds to salivary α-amylase with a cross-reactivity towards pancreatic α-amylase of less than 1% [7]. This antibody does not inhibit the salivary amylase activity. Hence two kinds of assay can be performed using this monoclonal antibody: an immunoassay monitoring the antibody-bound α-amylase activity, and an immunoprecipitation method, by which the salivary enzyme is removed from the serum and the activity of the pancreatic amylase is determined in the supernatant [7]. Very recently we learned of two groups reporting on monoclonal antibodies with similar properties [8, 9].

In this contribution we report on a convenient assay for pancreatic α-amylase in which the salivary amylase-binding antibody is attached to a solid support.

Fig. 1. Photograph of the antibody-loaded test strip, immersed into the serum sample.

Materials and Methods

The monoclonal antibody was prepared from a cell line, named 66C7, as described earlier [7]. This antibody was attached to a paper matrix [Gerber et al., to be published], resulting in a binding capacity of 1,100 mU salivary α-amylase per cm^2 of the paper. Test strips have been prepared, each with a 6×6 mm^2 piece of this paper. To bind the salivary α-amylase, 200 μl of human serum having a total α-amylase activity of less than 2,000 U/l is pipetted into a plastic tube (e.g. cat. No. 55.476, Sarstedt). Then a test strip bearing the antibody coated paper is placed into the test tube so that the 6×6 mm^2 paper field is immersed in the serum sample (fig. 1). The tube with the test strip and the sample is put on a shaker and shaken vigorously for 15 min at room temperature. The test strip, which now contains the bound salivary α-amylase is discarded and the residual amylase activity is determined in the supernatant.

The pancreatic isoamylase activity is calculated by the equation:

$$P = 1.15 \times R - 0.06 \times T,$$

where p = pancreatic α-amylase activity, R = residual activity and T = total activity.

The catalytic activity of α-amylase was determined at 37 °C by the *p*-nitrophenylmaltoheptaoside method (cat. No. 568 589, Boehringer Mannheim GmbH). The isoamy-

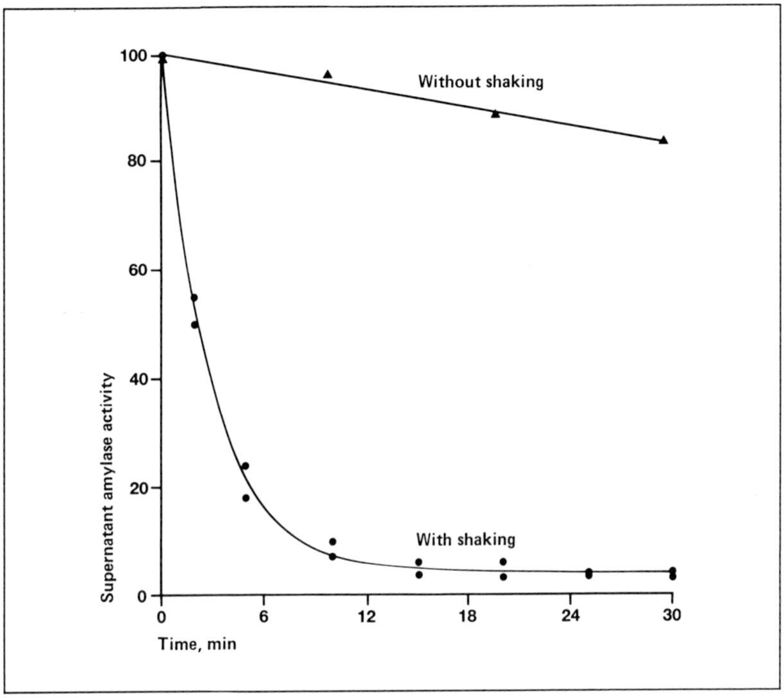

Fig. 2. Kinetics of the adsorption of salivary α-amylase to the antibody-coated test strip. 200 μl of a solution of salivary amylase (1,000 U/l) in PBS/BSA were treated as described under 'Materials and Methods'. After different times the supernatant was analysed for α-amylase activity (parallel experiments). Triangles = no movement; circles = vigorously shaken (Eppendorf laboratory shaker).

lase assay employing a wheat germ inhibitor from Boehringer Mannheim (cat. No. 779 601) was used for comparison. All other materials and methods are as described earlier [7].

Results and Discussion

As can be seen from figure 2, the adsorption of the salivary isoamylase to the antibody-coated surface of the test strip comes to a completion upon vigorous shaking after 15 min, whereas by diffusion alone, the adsorption reaction is very slow. At equilibrium 5% of the salivary

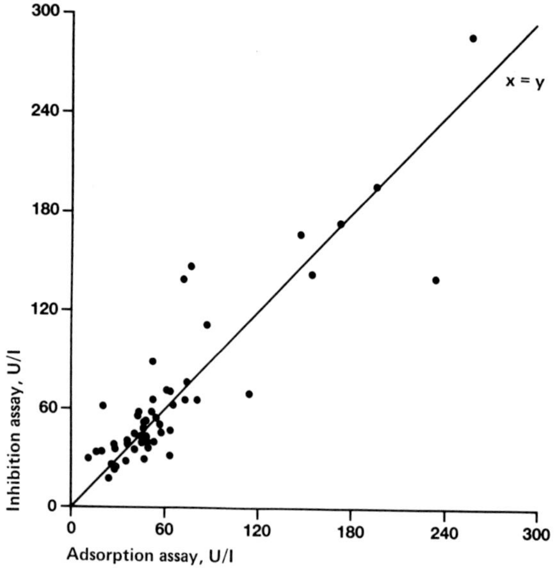

Fig. 3. Method comparison between the inhibition assay performed with wheat germ inhibitor and the immuno-adsorption assay. n = 30; r = 0.984.

enzyme remains unbound in the solution and 8% of the pancreatic isoamylase becomes unspecifically adsorbed to the matrix of the test strip. As had been shown earlier, the antibody does not bind the pancreatic amylase. Hence the true activity of the pancreatic isoamylase must be calculated using the equation given under 'Materials and Methods'. The coefficients of the equation vary slightly from batch to batch of the antibody-loaded paper.

The assay of pancreatic isoamylase using this technique has been compared with two other methods, the wheat-germ protein inhibition assay (fig. 3) and the enzyme immunoassay (fig. 4) as described earlier [7]. Both figures show that the methods correlate well with each other. The imprecision (CV) of the value for the pancreatic amylase as assayed by the adsorption assay is less than 5% within series.

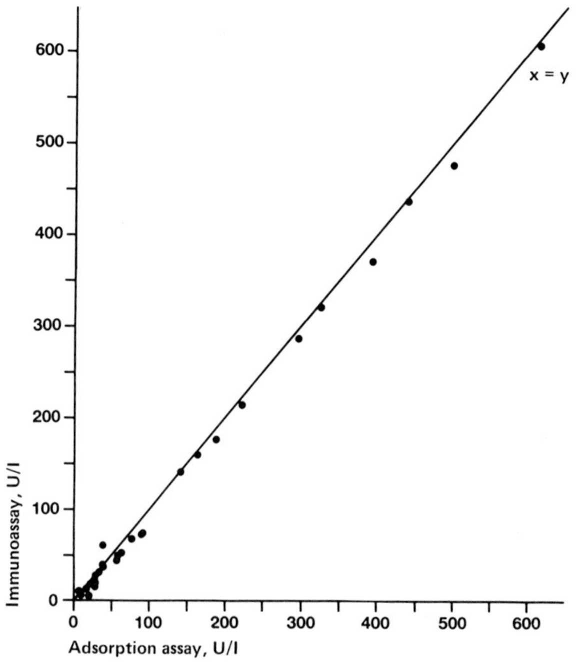

Fig. 4. Method comparison between the enzyme immunoassay as performed according to Gerber et al. [7] and the immuno-adsorption assay. n = 30; r = 0.998.

The number of antigenic sites by which the two α-amylase isoenzymes can be distinguished from each other must be very small due to the only minor differences in amino acid sequence. Simulation of the secondary structure according to the model of Chou and Fasman [10] showed three sites of immunogenicity between the amino acids 58 and 72, 175 and 190, and 316 and 375. Hence, we cannot exclude the possibility that monoclonal antibodies reported by other groups [8, 9] with similar properties as ours are directed against the same or a very close antigenic site.

References

1 Keller, P.J.; Kauffmann, D.L.; Allan, B.J.; Williams, B.L.: Further studies on the structural differences between the isoenzymes of human parotid α-amylase. Biochemistry *10:* 4867–4874 (1971).
2 Nakamura, Y.; Ogawa, M.; Nishide, T.; Emi, M.; Kosaki, G.; Himeno, S.; Matsubara, K.: Sequences of cDNAs for human salivary and pancreatic α-amylases. Gene *28:* 263–270 (1984).
3 Skude, G.: Electrophoretic separation, detection, and variation of amylase isoenzymes. Scand. J. clin. Lab. Invest. *35:* 41–47 (1975).
4 Royse, V.L.; Jensen, D.M.: Development of an agarose gel electrophoresis technique for determining α-amylase isoenzymes. Clin. Chem. *30:* 387–390 (1984).
5 Huang, W.Y.; Tietz, N.W.: Determinations of amylase isoenzymes in serum by use of a selective inhibitor. Clin. Chem. *28:* 1525–1527 (1982).
6 O'Connor, C.M.; McGeeney, K.F.: Isolation and characterization of four inhibitors from wheat flour which display differential inhibition specificities for human salivary and human pancreatic α-amylases. Biochim. biophys. Acta *658:* 387–396 (1981).
7 Gerber, M.; Naujoks, K.; Lenz, H.; Gerhardt, W.; Wulff, K.: Specific immunoassay of α-amylase isoenzymes in human serum. Clin. Chem. *31:* 1331–1334 (1985).
8 Ito, K.; Katayama, M.; Minamiura, N.; Yamamoto, T.: Preparation of human salivary α-amylase specific monoclonal antibody. J. Biochem. *97:* 1357–1362 (1985).
9 Mifflin, Th.E.; Benjamin, D.C.; Bruns, D.E.: Rapid quantitative, specific measurement of pancreatic amylase in serum with use of a monoclonal antibody. Clin. Chem. *31:* 1283–1288 (1985).
10 Chou, P.Y.; Fasman, G.D.: Prediction of the secondary structure of proteins from their amino acid sequence; in Meister, Advances in enzymology, vol. 47, pp. 45–148 (Wiley, New York 1978).

Dr. rer. nat. K. Wulff, Boehringer Mannheim GmbH, Bahnhofstrasse 9–15, D–8132 Tutzing (FRG)

Adv. clin. Enzymol. vol. 3, pp. 139–150 (Karger, Basel 1986)

Evidence for Multiple Forms of Pancreatic Lipase in Human Plasma

J.M. Lessinger[a], *P.L. Arzoglou*[b], *G. Férard*[a]

[a] Laboratoire du Centre de Traumatologie, Illkirch-Graffenstaden, France and Laboratoire de Biochimie Appliquée, Faculté de Pharmacie, Université Louis Pasteur, Strasbourg Cedex, France; [b] Laboratory of Biochemistry and Clinical Chemistry, Faculty of Chemistry, University of Thessaloniki, Greece

Intestinal absorption of fats requires a previous hydrolysis which is mostly achieved by enzymes of the pancreatic secretion. Pancreatic lipase (triacyl glycerol hydrolase EC 3.1.1.3) performs intraluminar degradation of triglycerides into free fatty acids and monoacylglycerol by heterogeneous catalysis. Determination of lipase activity in serum or plasma, as is the case for amylase activity, is now well established in biological diagnosis and monitoring of acute pancreatitis [1, 14]. Many authors consider the existence of isoenzymes to be a major inconvenience in the plasma amylase test, and lipase determination appears to have a better specificity, a higher analytical sensitivity and a longer duration of elevation in plasma [4, 8, 19].

Several pieces of evidence supporting the existence of multiple forms of pancreatic lipase have been reported. In 1963, Vogel and Zieve [21] noted that serum lipase from patients with acute pancreatitis has different optimal pH and bile salt requirement from those of lipase in normal serum. In 1971, Desnuelle [10] reported two distinct fractions of lipase activity after purification of porcine pancreatic juice on a Sephadex G-200 column. Moreover, two peaks of activity were also separated by chromatography on CM-cellulose of a pancreatic extract. In 1976, Kapaun et al. [17] reported, in one human plasma, two 'isolipases' separated by gel filtration. In 1977, De Caro et al. [9] separated two 'isolipases' from human pancreatic juice by polyacrylamide-gel electrophoresis, differing in their isoelectric points. The authors have not shown any difference in the antigenicity of these two forms.

We report here a body of evidence for at least two forms of lipase in human plasma.

Material and Methods

Lipase activity was determined with a nephelometer built in-house [12]. It has a laser source (λ = 632.8 nm), photodiodes positioned at 50°, 70° and 90° from the transmitted beam, and a thermostated compartment (37 °C) with a Peltier heating element. We measured light scatter at 50°, because the analytical sensitivity is better than at the other two angles. The signal (mV) was recorded continuously. In comparison with the initial reaction rate obtained for a purified preparation of swine pancreas lipase (previously titrated according to Rick [18]), lipase activity can be expressed in U/l (μmol of liberated fatty acids$\cdot min^{-1} \cdot l^{-1}$) [2].

A thermostated photometer (LKB Ultrolab System 8600) equipped with a recorder has also been used for the turbidimetric determination of lipase activity at 340 nm.

Commercially available swine pancreas lipase was used (Boehringer-Mannheim, Tutzing, FRG). Porcine colipase prepared as previously described [11] contained no lipase activity. Sodium deoxycholate was purchased from E. Merck, Darmstadt, FRG. An olive oil emulsion was prepared according to Tietz and Fiereck [20] and used at a final concentration of 0.25 g/l.

A commercially available substrate (Lipase UV system, Boehringer-Mannheim) has also been used [22]. It contained 0.30 mmol/l of triolein, 25 mmol/l of Tris-HCl buffer (pH 9.2), colipase at 3 mg/l, sodium deoxycholate at 18 mmol/l, and 0.1 mmol/l of $CaCl_2$.

We used the procedure of Tietz and Fiereck [20] with slight modifications [3], titrating the liberated fatty acids after 75 min incubation period at 37 °C with 50 mmol/l NaOH solution. A blank was run for each sample.

To compare the relative molecular masses (M_r) of lipase forms, gel filtration was performed on Sephadex G-100 (Pharmacia Fine Chemicals, Sweden). Two kinds of plasma (plasma from healthy subjects and plasma from subjects with acute pancreatitis) were applied to a column (1.14 × 75 cm) previously equilibrated at room temperature with Tris-HCl buffer 25 mmol/l (pH 8.8) containing 100 mmol of NaCl per litre. The same buffer was used for elution at a flow rate of 2.2 ml/h and fractions of 1.2 ml were collected. The elution volume corresponding to the tube containing the most lipase activity was considered to be the elution volume of lipase. The position of the peak of lipase activity was compared to that of albumin (assayed by bromocresol green method).

In order to show differences between the global molecular charges of the lipase forms, plasma samples from healthy subjects and diseased subjects were applied to a column of DEAE Trisacryl-M (IBF/Pointet-Girard, France).

The column (1.14 × 10 cm) was prepared according to the manufacturer's instructions and equilibrated at room temperature with Tris-HCl buffer 25 mmol/l (pH 8.0). The ions contained in the samples were previously eliminated by dialysis. The column was

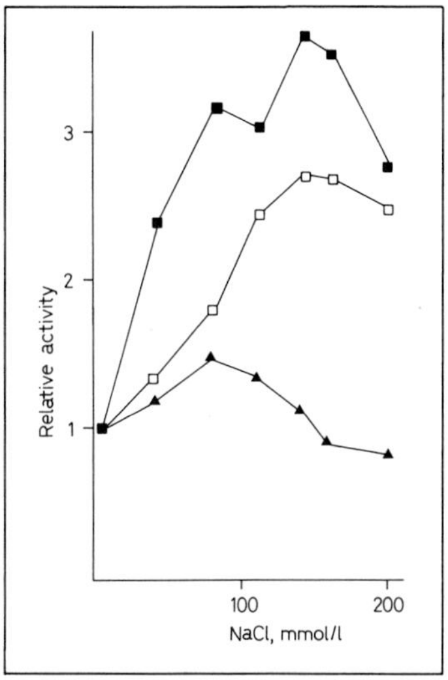

Fig. 1. Lipase activity as a function of NaCl concentration in the assay medium: ■ = pooled plasma from subjects with acute pancreatitis; ▲ = pooled plasma from healthy subjects; □ = purified swine pancreas lipase. Lipase activity was assayed by nephelometry using standard conditions (see 'Material and Methods'). Results (relative activities) are expressed as multiples of the activity without NaCl.

eluted with a continuous ionic strength gradient (NaCl dissolved in the same buffer), at a flow rate of 22 ml/h, and 1.2 ml fractions were collected.

Lipase concentration (μg/l) was also determined with a commercial kit (Enzygnost-Lipase, Behringwerke AG, Marburg, FRG) [7, 15]. It is an ELISA technique using tubes coated with antilipase antibodies produced against human pancreatic lipase from duodenal juice [13].

Results

Effect of NaCl Concentration

Using a commercially available reagent (UV System, Boehringer-Mannheim) containing colipase, we examined the effect of NaCl on the reaction rate with the nephelometric technique.

An increase of the reaction rate was noted for all the plasmas tested when NaCl was added to the assay medium at different concentrations (fig. 1). In plasmas from healthy subjects, a maximum of lipase activity was observed in the presence of NaCl 80 mmol/l. In contrast, lipase activity peaked in the presence of NaCl 140 mmol/l in plasmas from patients with acute pancreatitis. A similar pattern was also obtained in the case of purified swine pancreas lipase (fig. 1). The same effect was observed with a turbidimetric procedure.

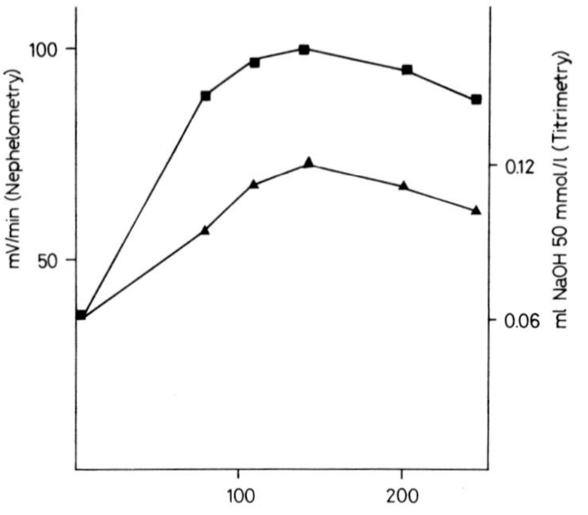

Fig. 2. Comparison of the effect of NaCl on lipase activity as determined by nephelometry (■) and titrimetry (▲). Results were obtained with an olive oil emulsion containing, per litre, 0.3 mmol of triolein, 3 mg of colipase, 18 mmol of deoxycholate and 25 mmol of Tris-HCl (pH 9.2).

Nephelometric or turbidimetric measurements were assessed by a titrimetric technique according to Tietz and Fiereck [20]. The activity of lipase was maximum in the presence of 140 mmol/l of NaCl for porcine pancreatic lipase when colipase (3 mg/l) was added in the assay medium (fig. 2), showing that NaCl increases the rate of release of fatty acids, indicating an activation effect of the salt.

The lipase-activating effect of NaCl at 140 mmol/l differed strongly between 50 adult plasmas tested. Degrees of activation ranged from 0.70- to 4.9-fold for lipase activity concentrations ranging from 51 to 1,625 U/l. Since the effect of NaCl might depend on the amount of enzyme in the sample, we studied the activation effect with aqueous dilutions of three different samples. As shown in figure 3, the extent of NaCl effect at 140 mmol/l did not depend on the dilution of the samples but on the kind of sample. In fact, pooled plasma from healthy subjects exhibited no lipase activation at 140 mmol/l whereas plasma from diseased subjects had a 3-fold increased activity.

To study the specificity of this effect, we replaced NaCl by KCl and by CH_3COONa at the same substance concentrations. The nature of the salts

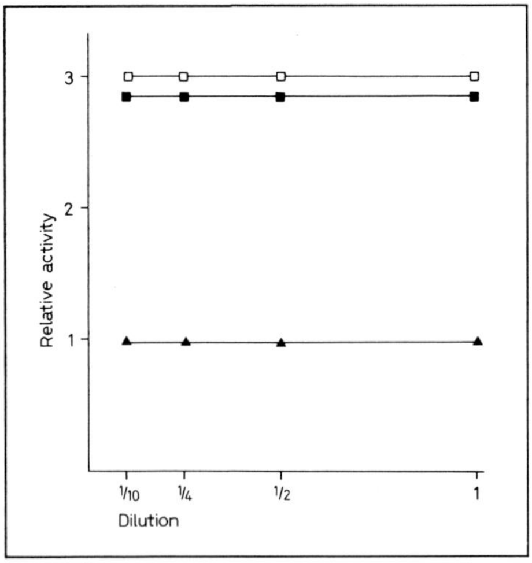

Fig. 3. Effect of sample dilution on the degree of activation of lipase by NaCl. ■ = Pooled plasma from subjects with acute pancreatitis; ▲ = pooled plasma from healthy subjects; □ = purified swine pancreas lipase. Relative activity is expressed as the ratio of activity measured with and without NaCl by nephelometry.

added to the assay medium did not change the observed patterns, indicating an effect of ionic strength. From those results, we concluded that human plasma might contain at least two forms of lipase which differed in their respective dependence to the ionic strength of the assay medium. In order to better characterize this property, we studied its relationship with colipase concentration.

Effect of Colipase Concentration on NaCl Effect

Two different lipase-containing preparations were tested for their sensitivity to NaCl concentration in the assay medium in the presence of various concentrations of colipase: plasmas from healthy subjects and from patients with acute pancreatitis (fig. 4).

In the absence of NaCl, there was no marked effect of colipase on plasma lipase activity from healthy subjects (fig. 4a) and lipase was moderately sensitive to colipase addition when pooled plasmas from diseased subjects were tested (fig. 4b). In the presence of NaCl (80 or 140 mmol/l), a strong inhibition was observed with all the plasmas studied when colipase was absent. Colipase concentrations exceeding 2 mg/l activated the lipase

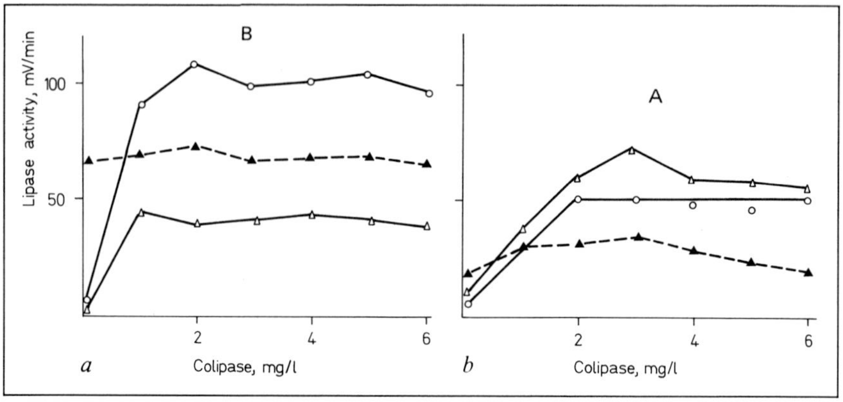

Fig. 4. Effect of colipase concentration on lipase activity of pooled plasma of healthy subjects *(a)* or of subjects suffering from acute pancreatitis *(b)*. Activity was determined by nephelometry in the absence of NaCl (▲), and in the presence of NaCl 80 mmol/l (○) or 140 mmol/l (△).

differently according to the kind of plasma tested in the presence of NaCl. Lipase activity from healthy subjects was always greater at 80 mmol/l of NaCl than at 140 mmol/l. Under these conditions, lipase activity was constant for colipase concentration between 2 and 6 mg/l (fig. 4a). In contrast, lipase activity from diseased subjects was always greater at 140 mmol/l NaCl than at 80 mmol/l NaCl. However, a maximum of activity was noted with 3 mg/l of colipase when the assay medium contained 140 mmol/l NaCl (fig. 4b). Thus, a colipase concentration of 3 mg/l may be retained for the assay of lipase (especially when NaCl at 140 mmol/l is added to the medium).

From these results, effect of ionic strength depends on colipase concentration. This is in keeping with results published by Borgstroem [5] and Borgstroem et al. [6] which showed that the ionic strength of the medium plays a major role in the interacting lipase-colipase-bile salts substrate. Furthermore, the effect of NaCl and colipase on lipase activity differed with the kind of plasma (healthy subjects or subjects with acute pancreatitis), confirming the existence of at least two forms of lipase in human plasma. One of them ('form A'), observed in plasma from subjects suffering from acute pancreatitis, can be defined by a greater activity with 140 than with 80 mmol/l NaCl. Another form present in healthy subjects ('form B') has an optimal activity in the presence of 80 mmol/l NaCl.

Purification of these two plasma forms was undertaken in order to confirm the existence of these two forms in human plasma, to see whether they can coexist in a same plasma, and to look for additional properties of these forms.

Separation of Forms A and B by Gel Filtration

When pooled plasma from healthy subjects was passed through a Sephadex G-100 column, a single peak of lipase activity was recovered with a relative mass (M_r) apparently exceeding 69,000 (M_r of human albumin). It corresponded to form B according to its NaCl concentration requirement for optimal activity (fig. 5a).

A peak in the same position was observed by lipase enzyme immunoassay (Enzygnost-Lipase, Behring), indicating that this enzymic form is recognized by human pancreatic lipase antibodies.

With the same column, pooled plasma collected on 6 consecutive days from a subject with acute pancreatitis gave a single peak of lipase activity, corresponding to a relative mass less than 69,000 (fig. 5b). This was also found in plasma from other patients with acute pancreatitis and for purified swine pancreas lipase. Activity was greater in the presence of 140 mmol/l of NaCl, indicating that this activity corresponded to form A, according to our definition. This form was also recognized by human pancreatic lipase antibodies. Thus, forms A and B differed by their relative masses, but both were recognized by human pancreatic lipase antibodies, indicating a partly common structure.

Separation of Forms A and B by Ion Exchange Chromatography

The most active fractions (25–29, see fig. 5b) following gel filtration of pooled plasma from the diseased subject were pooled. Two peaks were recovered when this pooled activity was passed through a column of DEAE Trisacryl-M at pH 8.0 (fig. 6). The major peak was not retained on the column, and the minor one, eluted with 50 mmol/l of chloride was of type B. Their respective positions indicated that the global charge of form B was more negative than that of form A at pH 8.0. These two peaks of lipase activity were recognized by human pancreatic lipase antibodies.

The most active fractions (19–25) obtained after gel filtration of pooled plasmas from healthy subjects (see fig. 5a) were also pooled and passed through the column of DEAE Trisacryl-M. They gave only one peak of lipase activity, of type B, recognized by the antilipase antibodies. The same results were obtained when a plasma sample from 1 subject without pancreatic disease was applied on the same column.

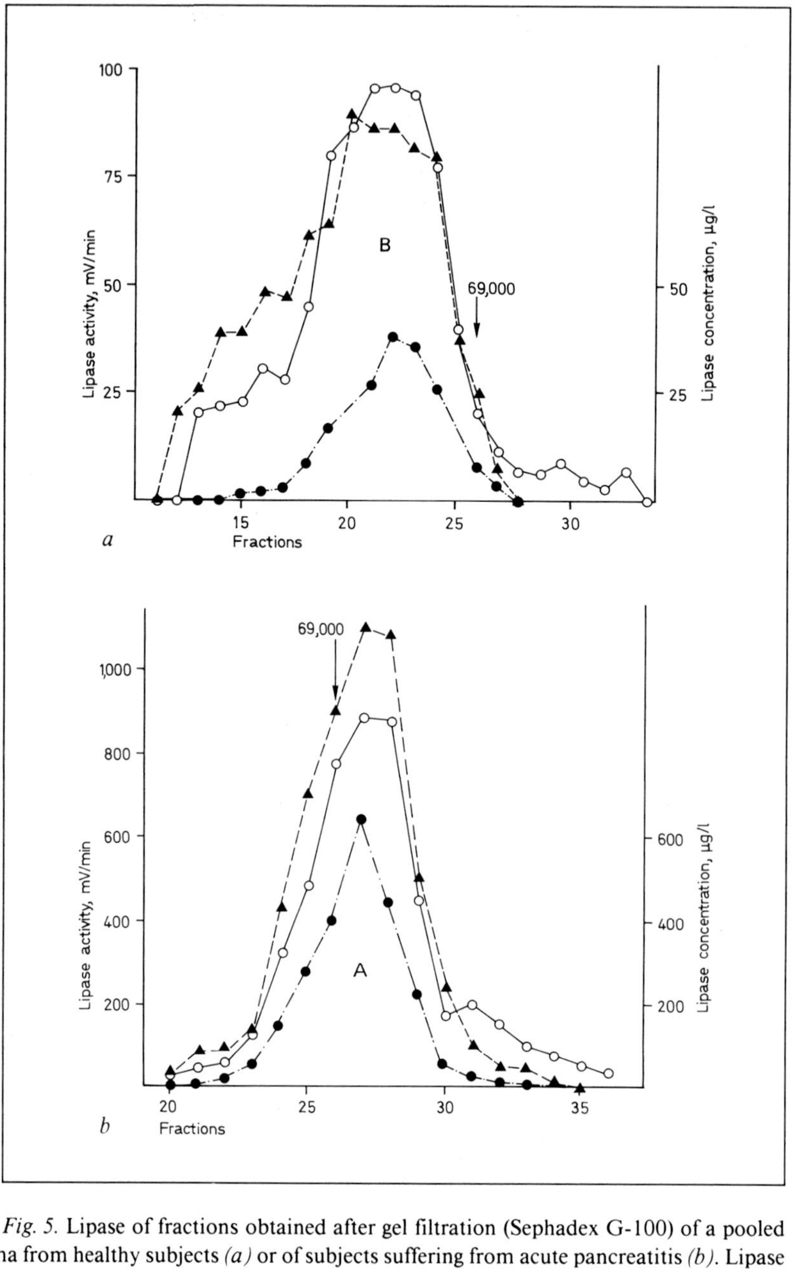

Fig. 5. Lipase of fractions obtained after gel filtration (Sephadex G-100) of a pooled plasma from healthy subjects *(a)* or of subjects suffering from acute pancreatitis *(b)*. Lipase activity was determined by nephelometry with NaCl 80 mmol/l (○) and with NaCl 140 mmol/l (▲). Lipase concentration was determined by enzyme immunoassay (●).

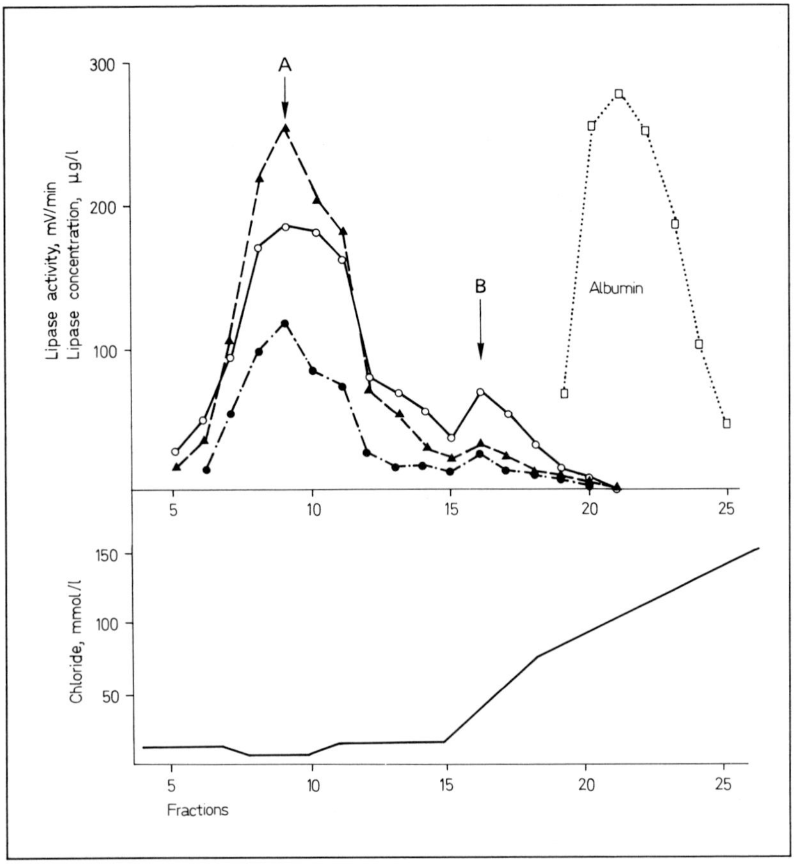

Fig. 6. Lipase of fractions obtained after ion-exchange chromatography (DEAE Tris-acryl-M, pH 8.0) of the pooled most active fractions (25–29; see fig. 5b). Lipase was determined by enzyme immunoassay (●), and by nephelometry with NaCl at 80 mmol/l (○), or at 140 mmol/l (▲).

Reactivity of Forms A and B towards Anti-Lipase Antibodies

Immunoreactivity of the two lipase forms was compared through their respective degree of immunoinhibition by human pancreatic lipase antibodies and by their avidity for these antibodies.

For immunoinhibition measurements 10 plasma samples (diluted 10-fold in water) were incubated for 2 h in plastic tubes coated with antilipase antibodies, then lipase activities were measured by nephelometry and the results were compared with those obtained for the same diluted samples incubated for 2 h in uncoated tubes (no antibodies). In the six samples from patients with acute pancreatitis, lipase activities (which corresponded to type A according to their NaCl concentration dependence) were strongly inhibited by this treatment (more than 60% of their initial activities). The 4 other plasma samples were from subjects with high lipase activity (when assayed without NaCl addition), low amylase activity and nonclinical symptoms of pancreatitis. Their lipase activities were of type B and were inhibited by antilipase antibodies by less than 31%.

The respective avidities of the two forms were deduced from results presented in figures 5 and 6 by calculating the lipase mass (when assayed by immunoassay and expressed in μg) per unit activity (when measured by nephelometry and expressed in U). This ratio was always ≥ 1 for form A and ≤ 0.7 for form B. Thus, forms A and B differ in the degree of immunoinhibition and in their respective avidities for human pancreatic lipase antibodies. This result confirms the presence of two different forms of lipase in human plasma with partially common structure. Furthermore, form A (which seems to be present only in plasma from subjects with pancreatic disease) appears to have a closer antigenic similarity with human pancreas lipase.

Additional Remarks and Conclusion

As lipase A was only observed in plasma from subjects suffering from acute pancreatitis, it is probably of pancreatic origin. Confirming this, lipase A was well recognized by human pancreatic lipase antibodies and we found some similar properties for both purified swine pancreas lipase and lipase that is released into the plasma during acute pancreatitis, especially in NaCl requirement.

The exact significance of lipase B is not known. Nevertheless, a pancreatic origin is also probable, since that form also splits triolein in the presence of colipase. In fact, Hoffmann and Weiss [16] have demonstrated that the addition of colipase (3 mg/l) to the assay medium suppresses the interference of lipoprotein lipase. Moreover, another lipase activity which may be present in human plasma is the so-called triglyceride lipase from hepatic tissue. Grenner et al. [15] have reported that human pancreatic lipase antibodies neither react with triglyceride lipase, nor with postheparin lipoprotein lipase; thus, these results and the recognition of form B by these antibodies are in favour of a pancreatic origin of this type of lipase.

In summary, a body of evidence supports the existence of at least two forms of lipase in human plasma. They are distinguished by their respective sensitivity to ionic strength and colipase concentration in the assay medium, relative molecular mass, global molecular charge, and reactivity with human pancreatic lipase antibodies. One form (lipase B) appears to be present in plasma from all subjects; the other one (lipase A) seems to be present only in plasma from patients with pancreatic disease. Some of the differentiating criteria could be the starting point of a study on current and even novel laboratory tests in order to improve specificity in diagnosis and monitoring of acute pancreatitis.

References

1 Amman, R.: Diagnose und Differentialdiagnose der Pancreatitis. Leber Magen Darm *6:* 217–226 (1976).

2 Arzoglou, P.L.; Férard, G.; Khalfa, K.; Métais, P.: Conditions for assay of pancreatic lipase from human plasma using a nephelometric technique. Clinica chim. Acta *119:* 329–335 (1982).

3 Arzoglou, P.L.; Férard, G.; Métais, P.: Differentiating two forms of plasma lipase by use of media with different ionic strength. Clin. Chem. *30:* 360–363 (1984).

4 Arzoglou, P.L.; Métais, P.; Férard, G.; Hollender, L.F.; Kauffmann, J.P.; Storck, D.: Intérêt de la détermination de la lipasémie au cours des affections pancréatiques. Comparaison avec le dosage de l'amylasémie. Presse méd. *12:* 273–275 (1983).

5 Borgstroem, B.: The action of bile salts and other detergents on pancreatic lipase and the interaction with colipase. Biochim. biophys. Acta *488:* 381–391 (1977).

6 Borgstroem, B.; Erlanson-Albertson, C.; Wieloch, T.: Pancreatic colipase. Chemistry and physiology. J. Lipid Res. *20:* 805–816 (1979).

7 Bowen, M.; Cooper, E.H.; McMahon, M.J.: The diagnosis of acute pancreatitis. Enhanced sensitivity from an ELISA assay of lipase. Biomed. Pharmacother. *37:* 395–398 (1983).

8 Courtois, P.; Art, G.; Vertongen, F.; Franckson, J.R.M.: Evolution of serum amylase, lipase and immunoreactive trypsin during pancreatitis attacks. Annls Biol. clin. *43:* 127–131 (1985).

9 De Caro, A.; Figarella, C.; Amic, J.; Michel, R.; Guy, O.: Human pancreatic lipase. A glycoprotein. Biochim. biophys. Acta *490:* 411–413 (1977).

10 Desnuelle, P.: La lipase pancréatique. Biochimie *53:* 841–852 (1971).

11 Erlanson, C.; Fernlund, P.; Borgstroem, B.: Purification and characterization of two proteins with colipase activity from porcine pancreas. Biochim. biophys. Acta *310:* 437–445 (1973).

12 Férard, G.; Thabet, M.; Schultz, G.; Grimm, P.; Métais, P.: Description d'un néphélomètre à source laser: application à la détermination d'une activité lipasique. Innov. Tech. Biol. méd. *2:* 53–65 (1981).

13 Forsell, E.: Purification and immunological quantification of human pancreatic lipase. Ann. Acad. Sci. Fenni. *166:* 5–36 (1974).

14 Fruchard, J.C.; Sezille, G.; Ghisbain, H.; Hassoun, A.; Choteau, P.; Arbon, G.; Quandalle, P.; Ribet, M.: Dosage de la lipase sérique: étude critique. Presse méd *9:* 509–512 (1980).

15 Grenner, G.; Deutsch, G.; Schmidtberger, R.; Dati, F.: Hochempfindlicher Enzymimmunoassay zur Bestimmung von Human-Pancreas-Lipase. J. clin. Chem. clin. Biochem. *20:* 515–519 (1982).

16 Hoffmann, G.E.; Weiss, L.: Specific serum pancreatic lipase determination with use of purified colipase. Clin. Chem. *26:* 1732–1733 (1980).

17 Kapaun, W.; Meier-Gabell, E.; Mueller-Wieland, K.; Berndt, W.: Verteilungsmuster von Lipase-Isoenzymen. Erste klinische Erfahrungen. Verh. dt. Ges. inn. Med. *82:* 963–965 (1976).

18 Rick, W.: Klinischer Test zur Bestimmung der Serum Lipaseaktivität. Z. klin. Chem. klin. Biochem. *5:* 530–539 (1969).

19 Roth, R.: Pancreatic enzymes other than amylase. Clin. Biochem. *12:* 272–274 (1979).

20 Tietz, N.W.; Fiereck, E.A.: Measurement of lipase activity in serum. Stand. Methods clin. Chem. *7:* 19–31 (1972).

21 Vogel, W.C.; Zieve, L.A.: A rapid and sensitive turbidimetric method for serum lipase based upon differences between the lipases of normal and pancreatitis serum. Clin. Chem. *9:* 168–181 (1963).

22 Ziegenhorn, J.; Neumann, U.; Knitsch, K.W.; Zwez, W.: Determination of serum lipase. Clin. Chem. *25:* 1067 (1979).

J.M. Lessinger, MD, Laboratoire de Biochimie Appliquée, Faculté de Pharmacie, Université Louis Pasteur, 74, route du Rhin, F-67048 Strasbourg Cedex (France)

Adv. clin. Enzymol., vol. 3, pp. 151–159 (Karger, Basel 1986)

Biochemical and Clinical Aspects of Isoenzymes of Glycolytic Enzymes in Plasma and Cerebrospinal Fluid

G.E.J. Staal

Department of Haematology, Division of Medical Enzymology,
Academic Hospital Utrecht, The Netherlands

Many enzymes exist in multiple forms. This may be caused by multiple genes, each encoding a different type of enzyme subunit, or by post-translational modification. The term isoenzyme should apply only to those multiple forms of enzymes arising from genetically determined differences in primary structure and not to those derived by modification from the same primary sequence (Commission on Biochemical Nomenclature).

Isoenzymes in serum or in cerebrospinal fluid are of clinical interest because they can be used as molecular markers of tissue damage. The diagnostic specificity of an enzyme test may be improved by measuring the activity of a particular isoenzyme rather than the total activity of an enzyme. However, the concentrations of most isoenzymes in serum or in cerebrospinal fluid are very low, and good approach is the use of isoenzyme-specific antisera.

Of the eleven enzymes involved in the conversion of glucose to lactate, at least seven are known to exist as isoenzymes (table I). Of these, lactate dehydrogenase is most often used in clinical diagnosis and several reviews on this subject have been published.

This paper considers two enzymes of the glycolytic chain: pyruvate kinase and enolase. Both exist as isoenzymes. The determination of the isoenzymes of enolase in serum (or plasma) and in cerebrospinal fluid has been proven to be of special importance in clinical diagnosis.

Table I. Isoenzymes of glycolytic enzymes

Enzyme	Isoenzyme type
Hexokinase	I, II, III and glucokinase
Phosphofructokinase	M, L and P
Aldolase	A, B and C
Phosphoglyceromutase	B and M
Enolase	α, β and γ
Pyruvate kinase	L, M and K
Lactate dehydrogenase	A, B and C

Pyruvate Kinase

Pyruvate kinase (ATP: pyruvate 2-0-phosphotransferase, EC 2.7.1.40) catalyses the conversion of phospho-enol-pyruvate to pyruvate with regeneration of ATP. At least three mammalian isoenzymes of pyruvate kinase are known, each of which is composed of four identical or nearly identical subunits. These isoenzymes are designated as L (liver), M (muscle) and K (kidney) [1]. The K type is also frequently designated as M_2. The subunit constitution of the isoenzymes can be designated by a subscript, e.g. M_4 for the tetrameric form of the M isoenzyme and K_2M_2 for a hybrid with two subunits of the K-type protomer and two of the M-type protomer.

The dominant isoenzyme in fetal tissue is type K. During development the other two appear. In adult organisms, type M is predominant in striated muscle and nervous tissue, while type L predominates in liver and erythrocytes. Type K is predominant in adult kidney and leucocytes. Table II shows the tissue localization and some properties of the various isoenzymes [2].

The M- and K-type isoenzymes are translated from different mRNAs [3]. The question of whether the M and K types are products of two distinct genes still remains to be elucidated. Genetic and biochemical findings, however, strongly suggest that the two different PK isoenzymes are under the control of only one structural gene [4].

The sequence of cloned K-type cDNA was found to be very similar to that of the M-type mRNA judging from results of cross-hybridization experiments [5]. This finding is consistent with the genetic finding that the M- and K-type mRNAs are transcribed from the same gene by differential

Table II. Tissue localization and properties of the various isoenzymes

Type	L	M	K
Occurrence	liver erythrocytes	brain heart muscle	leucocytes kidney intestine spleen
K_m PEP	0.5 m*M*	0.1 m*M*	1 to 2 m*M*
Alanine inhibition	moderate inhibition	no inhibition	strong inhibition

processing of a common nuclear RNA precursor or by gene rearrangements [5].

In view of these findings (one structural gene coding for the M and K type but two different mRNAs) the term isoenzyme as defined by the Commission on Biochemical Nomenclature, is now questionable.

It is well known that neoplasia may be associated with changes in isoenzyme distribution [6, 7]. A good example is the isoenzyme shift of pyruvate kinase in human brain tumours [8].

Type M_4 is predominant in adult brain whereas type K_4 is present in minor amounts. In fetal brain both K and M subunits are synthesized simultaneously. It seems likely that during development the synthesis of M subunits is favoured. Malignant gliomas are characterized by the absence of type M_4 whereas type K_4 and the hybrid K_3M are predominant. One can state cautiously that in gliomas the synthesis of fetal-type pyruvate kinase is favoured whereas the synthesis of the M type (predominant in the tissue of origin) is depressed [8].

A good correlation also exists between the amount of K type and degree of differentiation of primary brain tumours [9]: in the highly malignant glioblastoma K_4 type is predominant, while in well-differentiated astrocytomas grade I and grade II M_4 is also expressed besides K_4, but is still different from the normal brain tissue (fig. 1).

One may expect that damage within the central nervous system should result in raised values of intracellular products within the cerebrospinal fluid. This is true for the enzyme enolase but not for pyruvate kinase. This is probably caused by binding of pyruvate kinase to actin which prevents its release.

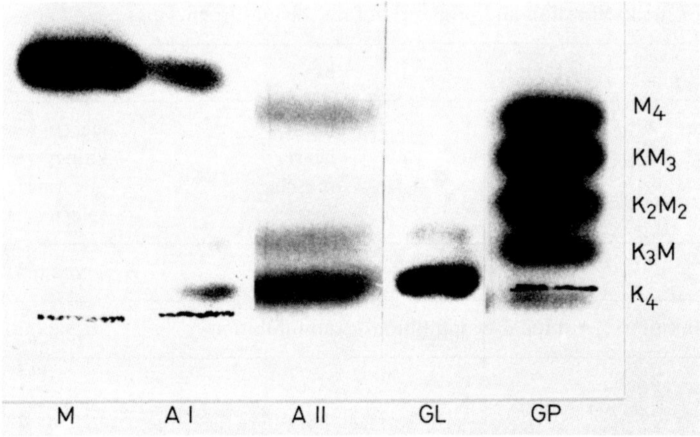

Fig. 1. Electrophoretic pattern of pyruvate kinase of primary brain tumours. A I = astrocytoma grade I; A II = astrocytoma grade II; GL = glioblastoma; M = muscle; GP = glandula pinealis served as a reference [from ref. 9].

Pyruvate Kinase in Plasma

The measurement of pyruvate kinase activity in serum or plasma has been used to identify carriers of Duchenne's muscular dystrophy and patients with progressive muscular dystrophy, and to diagnose the extent of myocardial infarcts [10, 11]. Kinetic and electrophoretic studies indicated that the pyruvate kinase isoenzyme found in serum from affected patients is mainly M-type pyruvate kinase, which is the only pyruvate kinase found in skeletal muscle and is the major component in myocardium. It has been reported that normal serum or plasma contains a mixture of the L and the M isoenzymes of pyruvate kinase [12]. However, Ibsen [1] clearly demonstrated the presence of both K and M isoenzymes; the latter predominates. In conclusion, these studies suggest that serum pyruvate kinase determination may be of interest for certain types of neuromuscular diseases and therefore of critical importance in the clinical diagnosis of affected patients.

Enolase

Enolase (EC 4.2.1.11) catalyses the interconversion of 2-phospho-*D*-glycerate and phospho-enolpyruvate in the glycolytic pathway. Enolase is a dimeric enzyme consisting of two subunits coded for by three loci: α, β and

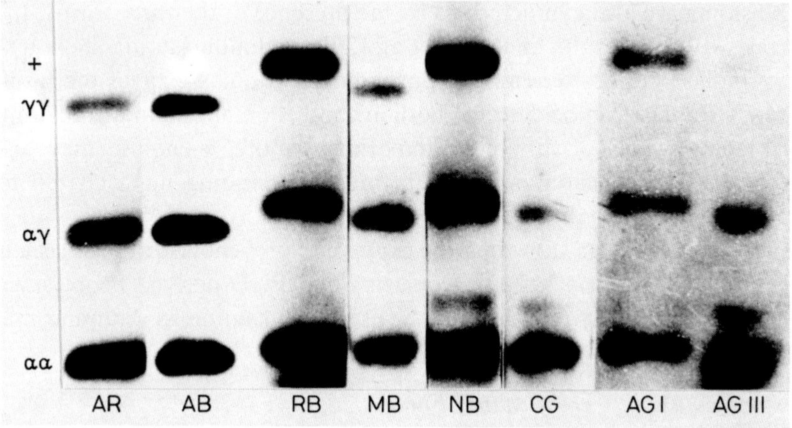

Fig. 2. Electrophoretic pattern of enolase of different kinds of tumours. AR = Adult retina; AB = adult brain; RB = retinoblastoma; MB = medulloblastoma; NB = neuroblastoma; CG = child glioma; AG I = astrocytoma grade I; AG III = astrocytoma grade III.

γ [14]. In most adult tissues only the αα dimer is present. Adult muscle and heart contain the ββ dimer; in the heart αβ hybrids are found as well. The γ-subunit was previously believed to be expressed exclusively in mature neurons [15] which resulted in the designation of γγ-enolase as neuron-specific enolase (NSE). Apart from being confined solely to neurons within the central nervous system, NSE was subsequently found in peripheral neurons and ganglia [16].

It has been reported [17] that peripheral neuroendocrine cells (APUD cells) contain NSE in addition to non-neuronal enolase, thus giving rise to three isoenzymes (αα, αγ and γγ). Tumours arising from these APUD cells were also shown to contain NSE, which resulted in its use as a marker for tumours of neuroendocrine origin [18].

Neuron-Specific Enolase

The finding that γ-enolase is essentially neuronal and that astrocytes stain strongly for α-enolase suggested the use of these isoenzymes as markers for neuro-ectodermal tumours. However, recent findings have demonstrated that neuron-specific enolase in neoplasia is not restricted to nervous system neoplasms or APUD-derived neoplasms [19]. Figure 2 shows the electrophoresis of enolase in different kinds of tumours. The α and γ subunit is expressed in normal brain and retina. Retinoblastoma and neuroblastoma are also characterized by the presence of the three forms. In contrast with the results of Royds et al. [20], medulloblastoma show αγ- and γγ-enolase besides αα-enolase. Surprisingly, gliomas express the γ-subunit too. γ-Enolase can be detected both in child gliomas and in astrocytomas of different grades of differentiation. Furthermore, γ-enolase may also be found in breast carcinoma and in breast carcinoma metastasized to the brain.

These findings show that the expression of γ-enolase in neoplasia is not restricted to neuron system neoplasms or APUD-derived neoplasms, and demonstrate that γ-enolase must be used with caution as a tumour marker.

Enolase in Cerebrospinal Fluid

Royds et al. [21] proposed that enolase isoenzymes are very sensitive markers of neurological disease and that assay of the α- and γ-isoenzymes appears to reflect damage to the glial and neuronal cells, respectively. They studied both α- and γ-enolase in the cerebrospinal fluid in a variety of neurological disorders including herpes simplex, encephalitis, migraine, epilepsy, dementia, multiple sclerosis and various types of neoplasms.

Cerebrospinal fluid α-enolase was found to be raised in most cases studied. γ-Enolase on the other hand was raised in fewer patients and to a lesser extent than α-enolase. α-Enolase was raised in 23 out of 28 cases of primary brain tumours and in all 3 cases of secondary tumours studied. In contrast, only 18% of the cases with primary tumours showed any rise in γ-enolase but this isoenzyme was elevated in all 3 cases with metastatic tumours. These findings indicate greater frequency of neuronal damage associated with secondary tumours of the brain. Preliminary results in our own laboratory, using a Pharmacia RIA 100 kit for estimation of γ-enolase, indicated no rise of γ-enolase in cerebrospinal fluid of patients with multiple sclerosis and in most cases of patients with primary brain tumours.

Interesting in this connection are the results of Royds et al. [22] in patients with acute lymphoblastic leukaemia treated with chemotherapy. Cerebrospinal fluid γ-enolase was found to be raised in the majority of cases studied, indicating neuronal damage.

Enolase in Serum or Plasma

Human neuroblastoma cells can express γ-enolase, as was first demonstrated by Dhillar et al. [23]. Serum levels of γ-enolase in children with neuroblastoma have been studied by Zeltzer et al. [24] in a series of 18 patients with metastatic neuroblastoma. The serum enolase was raised from 7.5 to 207 ng/ml. Values above 100 ng/ml indicated poor prognosis.

The potential use of plasma levels of γ-enolase in the diagnosis and treatment of melanocytic lesions was demonstrated by Taylor et al. [25] who determined levels of this isoenzyme in 20 patients with histologically confirmed lesions. Plasma γ-enolase was elevated in 9 patients with cutaneous malignant melanoma and in all of the 8 patients with disseminated malignant melanoma. In 5 out of 6 patients in whom pre- and post-operative values were studied, the plasma γ-enolase levels fell after resection of the melanoma and in one of these relapse after 12 months was accompanied by a rise of γ-enolase.

Neuron-specific enolase is elevated in a majority of patients with small-cell lung cancer, especially those with extensive disease [26]. In a study of 93 patients with small-cell lung cancer it was found [26] that neuron-specific enolase levels tended to correlate with tumour burden; the highest values were found in patients with metastatic disease while lower values occurred in those with tumour confined to the chest. Furthermore, γ-enolase levels declined toward normal in patients demonstrating clinical response to chemotherapy and rose in relapsing patients [26]. Similar results were reported by Arioshi et al. [27] and Carney et al. [28].

In summary, enolase activity in the cerebrospinal fluids is a sensitive indicator of pathological change. The levels of γ- and α-enolase may reflect the degree of involvement of neurons and non-neuronal cells, respectively. There appears to be a useful relationship between isoenzyme levels and the extent of tissue damage.

In patients with known forms of neoplasia such as small cell carcinoma of the lung, melanoma and neuroblastoma, serum γ-enolase is a useful indicator of tumour load, progress of the disease and response to treatment.

References

1 Ibsen, K.H.: Interrelationships and function of the pyruvate kinase isozymes and their variant forms. A review. Cancer Res. *37:* 342–353 (1979).

2 Staal, G.E.J.; Rijksen, G.: Regulation of pyruvate kinase in normal and pathological conditions; in Beitner, Regulation of carbohydrate metabolism, vol. I, pp. 143–159 (CRC Press, Boca Raton 1985).

3 Noguchi, T.; Tanaka, T.: The M_1 and M_2 subunits of rat pyruvate kinase are encoded by different messenger RNAs. J. biol. Chem. *257:* 1110–1113 (1982).

4 Saheki, S.; Saheki, K.; Tanaka, T.; Tanaka, T.: Peptide structures of pyruvate kinase isozymes. II. Origins of types M_1 and M_2 isozymes suggested from species-variations in their peptide maps. Biochim. biophys. Acta *704:* 494–502 (1982).
5 Imamura, K.; Noguchi, T.; Tanaka, T.: Regulation of isozyme patterns of pyruvate kinase in normal and neoplastic tissues; in Staal, van Veelen, Markers of human neuroectodermal tumours (CRC Press, Boca Raton, in press).
6 Weinhouse, S.: New dimensions in the biology of cancer. Cancer *45:* 2975–2980 (1980).
7 Weber, G.: Enzymology of cancer cells. New Engl. J. Med. *296:* 486–493 (1977).
8 Van Veelen, C.W.M.; Verbiest, H.; Vlug, A.M.C.; Rijksen, G.; Staal, G.E.J.: Isozymes of pyruvate kinase from human brain, meningiomas and malignant gliomas. Cancer Res. *38:* 4681–4687 (1978).
9 Van Veelen, C.W.M.; Verbiest, G.; Zülch, H.; Ketel, B.A. van; Vlist, M.J.M. van der; Vlug, A.M.C.; Rijksen, G.; Staal, G.E.J.: *L*-α-Alanine inhibition of pyruvate kinase from tumors of the human central nervous system. Cancer Res. *39:* 4263–4269 (1979).
10 Sage, J.; Inati, Y.; Samaka, F.: The importance of serum pyruvate kinase in neuromuscular diseases and carrier states. Muscle Nerve *2:* 390–393 (1979).
11 Percy, M.E.; Chang, L.S.; Murphy, E.G.: Serum creatine kinase and pyruvate kinase in Duchenne muscular dystrophy carrier detection. Muscle Nerve *2:* 329–339 (1979).
12 Edwards, R.J.; Watts, D.C.: Specific spectrophotometric assay for the M isoenzyme of pyruvate kinase in plasma samples containing mixtures of the muscle (M) and liver (L) isoenzymes. Clin. Chem. *27:* 906–909 (1981).
13 Lani Ngo, J.; Garratt, K.N.; Eversole, P.E.; Orlando, R.A.; Ibsen, K.H.: Pyruvate kinase isoenzymes in human serum. Clin. Chem. *30:* 111–113 (1984).
14 Pearce, J.M.; Edwards, Y.H.; Harris, H.: Human enolase isozymes: Electrophoretic and biochemical evidence for three loci. Ann. hum. Genet. *39:* 263–276 (1976).
15 Marangos, P.J.; Zis, A.P.; Clark, R.L.; Goodwin, F.K.: Neuronal, non-neuronal and hybrid forms of enolase in brain. Structural, immunological and functional comparison. Brain Res. *150:* 117–133 (1978).
16 Marangos, P.J.; Schmechel, D.E.; Parma, A.M.; Clark, R.L.; Goodwin, F.K.: Measurement of neuron-specific (NSE) and non-neuronal (NNE) isozymes of enolase in rat, monkey and human nervous tissue. J. Neurochem. *33:* 319–329 (1979).
17 Marangos, P.J.; Polak, J.M.; Pearse, A.G.E.: Neuron specific enolase. A probe for neurons and neuroendocrine cells. Trends Neurosci. *5:* 193–195 (1982).
18 Tapia, P.J.; Polak, J.M.; Barbosa, A.J.A.: Neuron-specific enolase is produced by neuro-endocrine tumours. Lancet *i:* 808–811 (1981).
19 Vinores, S.A.; Bonnin, J.M.; Rubinstein, L.J.; Marangos, P.J.: Immunohistochemical demonstration of neuron-specific enolase in neoplasms of the CNS and other tissues. Archs Path. Med. *108:* 536–540 (1984).
20 Royds, J.A.; Parsons, M.A.; Taylor, C.B.; Timperley, W.R.: Enolase isoenzyme distribution in the human brain and its tumours. J. Path. *137:* 37–49 (1982).
21 Royds, J.A.; Davies-Jones, G.B.; Lewtas, N.A.; Timperley, W.R.; Taylor, C.B.: Enolase isoenzymes in the cerebrospinal fluid of patients with diseases of the nervous system. J. Neurol. Neurosurg. Psychiat. *46:* 1031–1036 (1983).

22 Royds, J.A.; Lilleyman, J.S.; Timperley, W.R.; Taylor, C.B.: Cerebrospinal fluid enolase isoenzymes and neurotoxicity in early treatment of lymphoblastic leukemia. Archs Dis. Childh. *59:* 267–271 (1984).
23 Dhillon, A.P.; Rode, J.; Leatham, A.: Neuron-specific enolase: an aid to the diagnosis of melanoma and neuroblastoma. Histopathology *6:* 81–92 (1982).
24 Zelber, P.M.; Marangos, P.J.; Parma, A.M.; Sather, H.; Dalton, A.; Hammond, D.; Siegel, S.E.; Seeger, R.C.: Raised neuron-specific enolase in serum of children with metastatic neuroblastoma. Lancet *ii:* 361–363 (1983).
25 Taylor, C.B.; Royds, J.A.; Timperley, W.R.: Alpha and gamma enolase in the assessment of tumours of neuroectodermal origin; in Staal, van Veelen, Markers of human neuroectodermal tumours (CRC Press, Boca Raton, in press).
26 Johnson, D.H.; Marangos, P.J.; Forbes, J.T.; Hainsworth, J.D.; Van Welch, R.; Hande, K.R.; Greco, F.A.: Potential utility of serum neuron-specific enolase levels in small cell carcinoma of the lung. Cancer Res. *44:* 5409–5414 (1984).
27 Ariydoshi, Y.; Kato, K.; Ishiguro, Y.; Ota, K.; Sata, T.; Suchi, T.: Evaluation of serum neuron-specific enolase as a tumor marker for carcinoma of the lung. Gann *74:* 219–225 (1983).
28 Carney, D.N.; Marangos, P.J.; Ihde, D.C.; Bunn, P.A.; Cohen, M.H.; Minna, J.D.; Gazdar, A.F.: Serum neuron-specific enolase: a marker for disease extent and response to therapy of small-cell lung cancer. Lancet *i:* 583–585 (1982).

G.E.J. Staal, MD, Department of Haematology, Division of Medical Enzymology, Academic Hospital Utrecht, P.O. Box 16250, NL–3500 GC Utrecht
(The Netherlands)

Adv. clin. Enzymol., vol. 3, pp. 160–167 (Karger, Basel 1986)

A Marked Increase of Lactate Dehydrogenase Isoenzyme Three

R.N.M. Weijers, R.P.J. Oude Elferink, J. Mulder, H. Kruijswijk

Department of Clinical Chemistry, Onze Lieve Vrouwe Gasthuis, Amsterdam, The Netherlands

That an enzyme is involved in the conversion of pyruvate to lactate has been known since 1932 [1]. In the past 53 years considerable advances have been achieved in the knowledge of lactate dehydrogenase (LD; EC 1.1.1.27), a tetrameric molecule in which two different polypeptides may participate as subunits; a molecule with a 222 symmetry and a relative molecular mass (M_r) of 140,000.

Already in 1967, *Ganrot* [2] presented evidence of a macro LD. His study indicated that the abnormal LD pattern was due to complexing between LD and the immunoglobulin A (IgA). At the present time several studies have confirmed these findings. Analysis of all these publications [2–21] leads to the conclusion that two main types of macromolecular LD exist, namely NAD^+-dissociable macromolecular LD, and non-NAD^+-dissociable macromolecular LD. The former group comprised 7 cases, 2 presenting signs of an autoimmune disease. The latter group consisted of 48 reported cases who were apparently healthy patients. This group may be subdivided into group G, which comprised 13 cases whose macro LD contained the immunoglobulin of the class G, and group A, which comprised 35 cases whose macro LD contained the immunoglobulin of the class A. Because of a lack of cases of subgroup G, our attention has been focused on subgroup A. Regardless of the electrophoretic techniques used, such as agarose-gel electrophoresis, agar-gel electrophoresis or cellulose acetate electrophoresis, the complexes of this subgroup are associated with the following abnormal LD isoenzyme patterns: a marked increase of activity of the LD-3 band (n = 17); the absence of activity of the LD-2 band accompanied by an

Table I. Improved purification procedure of macro LD-IgA_{κ} from human plasma

	Volume ml	Recovery %	Purification factor	Specific activity kU/g
Old method	48	28.1	10,775	57.0
New method	945	61.8	104,000	139.3

additional band of LD activity (n = 15); a shift of the electrophoretic mobility of the bands of LD activity (n = 1); or the presence of a broad diffuse zone of LD activity (n = 2). Since 1981, in cooperation with Dr. Mulder, Clinical Immunology, Laboratory of the Public Health organization of the Frisian province, and Dr. Oude Elferink of the section for medical enzymology and metabolism of the Academic Medical Center of the University of Amsterdam, we have investigated the marked increase of LD-3.

The complex was isolated in an almost pure form by gel filtration and affinity chromatography. We originally needed 3 weeks to purify the complex starting from 48 ml of serum with a recovery of 28%. In order to obtain sufficient pure macro LD, we have changed and scaled up the procedure and today we are able to isolate the complex in the same time interval starting from 945 ml of plasma, with an improved recovery, an increased purification factor, and a higher figure for the specific activity (table I). The estimated protein content ratio of IgA to purified complex based on the results of the new procedure is 0.40, a figure correlating fairly well with the theoretical value of 0.38 for a complex composed of one molecule of monomeric IgA associated with two molecules of tetrameric LD. This result, therefore, supports the structure of this complex already proposed.

The immunoglobulin subunit of the complex is obtained by dissociation of the pure complex with 3 *M* NH_4SCN followed by Sephacryl S300 gel filtration. The final preparation has a M_r of approximately 180,000 as determined by gel filtration, and unchanged immunochemical proporties as showed by immunoelectrophoresis in agarose gel containing monospecific IgA antiserum.

In order to determine the nature of the LD isoenzyme of the complex, the 5 different LD isoenzymes were separately purified from the serum of a patient by affinity chromatography and preparative thin-layer electrophoresis. A solution of the pure IgA was incubated at room temperature with a

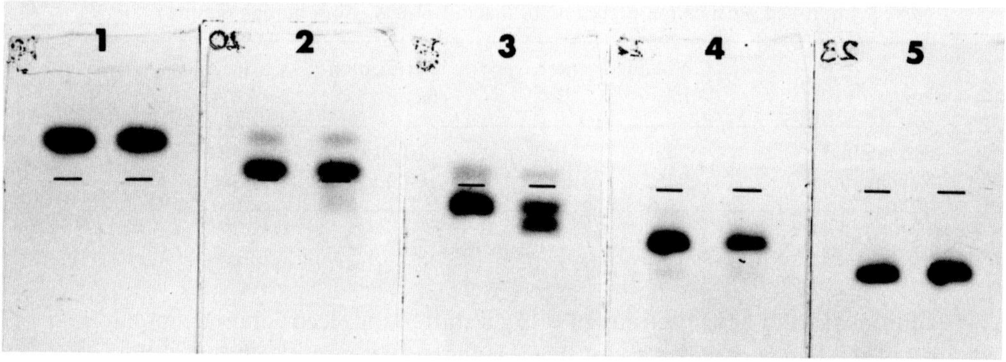

Fig. 1. Agar-agarose LD patterns of purified LD isoenzymes 1 through 5 before (left) and after (right) incubation with IgA_κ purified from LD-IgA_κ complex.

solution of a purified LD isoenzyme, subsequently separated electrophoretically on agar-agarose gel, and stained for LD activity. In figure 1, the purified LD isoenzyme preparation was applied on the left of each slide numbered 1 through 5. The isoenzyme number corresponds with the number of the slide. The same sample incubated with purified IgA_κ was applied on the right side. It is obvious that only LD-3 has been involved with the correct recombination to the original complex as demonstrated by the remarkable similarity of the electrophoretic mobility between the purified native and reassembled complex. An accurate examination of electrophoretic pattern number 2 reveals some 'smeared' LD activity, an indication of a weak interaction between the IgA_κ and LD-2.

To verify these results, original purified complex was adsorbed to polystyrene microtitre plates. After emptying the plate and exhaustively washing, glycine-HCl buffer of pH 3.0 was added to dissociate the complex into its subunits. After several washing cycles purified LD isoenzyme, containing 0.05% Tween, was added and allowed to react. Unbound isoenzyme was removed and a solution of LD substrate containing NAD^+ added. The conversion of NAD^+ to NADH was detected by a bioluminescence assay. From these results (fig. 2) it is seen that LD-3 is essential for obtaining original complex LD activity. The results shown by subset 2 sustained the experimental evidence of a minimal interaction between IgA_κ and LD-2, as demonstrated by the electrophoretic pattern.

It should be mentioned that the binding between IgA_κ and LD-3 can be blocked by anti-light chain kappa antiserum and not by anti-Fc chain IgA

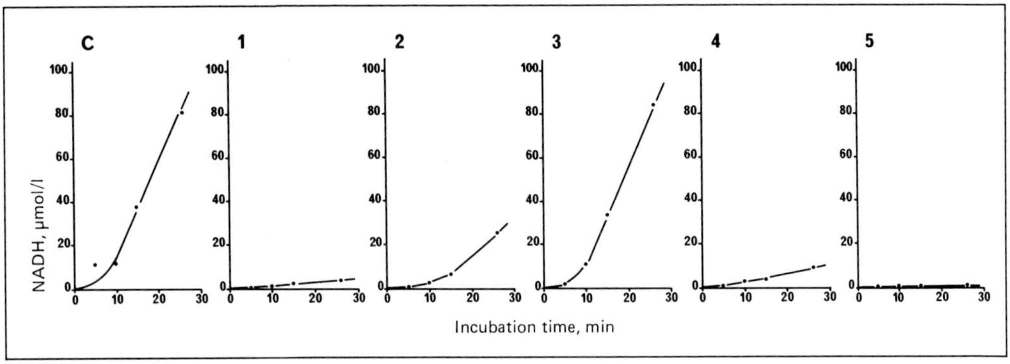

Fig. 2. Conversion of NAD⁺ to NADH by LD linked to IgA_κ. The IgA_κ originates from native, purified macro LD adsorbed to polystyrene. C, LD present in native macro LD; 1 through 5, LD-1, LD-2, LD-3, LD-4, and LD-5 present after glycine-HCl dissociation of the native macro LD, followed by incubation of the remaining IgA_κ with LD-1, LD-2, LD-3, LD-4, and LD-5, respectively.

antiserum, indicating that the binding of LD-3 occurs at the Fab region of the immunoglobulin.

From experimental results obtained with LD isoenzymes purified from the erythrocytes of a patient as well as from a human serum pool, it can be concluded that the IgA_κ recombines not only with autologous LD-3 but also with homologous LD-3. These results indicate that the formation of macro LD is due to the occurrence of specific LD-3 binding IgA_κ molecules and not to alterations in the patient's LD-3.

It is tempting to speculate that this IgA_κ is the expression of a viral antigen, a bacterial antigen or another infectious exogenous antigen that exhibits a unique cross-reactivity with LD-3 due to comparable structures of closely allied determinants, or the coïncidental fitting of the completely non-allied structure of LD-3 on the epitope of this monoclonal antibody.

At the beginning of this investigation much attention was paid to the enzyme part of the complex. At the present time we have shifted our attention to the immunoglobulin part. From experimental results it can be concluded that free cross-reacting IgA_κ is not demonstrable in the patient's serum. In order to determine the serum concentration of the complexed cross-reacting immunoglobulin, we separated in 5 cases, the macro LD from the low molecular weight LD by AcA 22 gel-filtration chromatography. Since 1 U of complexed LD activity equals 2.68 µg of cross-reacting

Table II. Properties of type 1 macro CK and type 1 macro LD

	Macro CK	Macro LD
Immunoglobulin nature	IgG	IgA
Specificity in binding antigen	CK-BB	LD-3
Limited binding capacity	+	1 μg = 0.37 U
Antigen-binding region	Fab	Fab
Type of antibody	?	monoclonal
Type of light chain	λ/κ	κ
Molar ratio Ig:enzyme	1:2	1:2
Molar ratio $Ig^{X}_{comp.}:Ig^{X}_{total}$	1:9,000	1:10,000
Binding of subunits	reversible	reversible
Enzyme composition	unaltered	unaltered
Stability	high	high
Observed persistence	> 2 years	> 6 years
Frequency	1:1,000	1:20,000

IgA_{κ}, the range of cross-reacting IgA_{κ} varies in this group from 236 to 1,144 μg/l serum. The main conclusion of these observations is that a profound difference exists in the individual immune response.

Most autoantibodies develop without apparent cause, fluctuate inexplicably and either disappear enigmatically or persist indefinitely as do the LD complexes. A single mechanism could hardly explain these phenomena, nor could a single aberration account for the many varieties of autoimmunization. Recent investigations have emphasized two mechanisms that can lead to the production of autoantibodies: impaired immunoregulation, which causes the formation of autoantibodies against a family of ubiquitous autoantigens, and the idiotype cascade which can foster the production of autoantibodies by means of loopholes in the rules against autoimmunization. A final answer to these questions is far away and, for instance, must await the characterization and quantitation of the number of lymphocyte subpopulations present in the blood of these individuals, a process currently underway in our group.

The properties of the macro LD complex may be compared with those of macro creatine kinase (CK; EC 2.7.3.2) type 1. These data (table II) are based on the published results of Jockers-Wretow and Plessing [22], Stein et al. [23–27], Urdal [28], Lang and Würzberg [29], and Wu et al. [30].

It is evident that a subunit of both complexes belongs to one of the major classes of immunoglobulin. A high specificity in the binding of antigen has been demonstrated, and a limited and well-defined antigen-binding capacity has been established. The binding of the antigen occurs in the antigen-binding region (the Fab region) of the immunoglobulin. From these results it can be concluded that the requirements are fulfilled for calling this subunit an antibody; furthermore, the results of high resolution electrophoresis show that the IgA_{κ} is a monoclonal antibody, lambda as well as kappa light chains have been identified as a subunit of the macro CK type 1 complex, whereas kappa light chains have been identified in the majority of LD-IgA complexes. In both of the complexes the isoenzyme is bound in equal molecular ratios onto the Fab sites of the immunoglobulin, indicating 1 in 2 complexes. The complexing capacity of the purified immunoglobulin has been found to be about 1 in 10,000; thus, 1 out of 10,000 molecules of the patient's IgG and IgA is able to bind one molecule CK-BB and one molecule LD-3, respectively. This ratio may, however, vary from case to case. Some authors report that the binding between the enzyme and the immunoglobulin is reversible and not due to alterations in the molecular enzyme composition. Both complexes show a high stability and persist for long periodes of time. The macro CK has been found in about 1 in 1,000 cases, whereas our figures on the incidence of the LD-IgA_{κ} complex indicate a frequency of about 1 in 20,000 cases.

There is clearly a strong analogy between the two complexes. Therefore, I propose to adopt the generally accepted nomenclature of macro CK type 1 in a slightly modified form for this LD complex and to term the complex: macro LD type 1.

References

1 Banga, I.; Szent-Györgyi, A.; Vargha, L.: Über das Co-Ferment der Milchsäure Oxydation. Hoppe-Seyler's Z. physiol. Chem. *210:* 228–235 (1932).

2 Ganrot, P.O.: Lupoid cirrhosis with serum lactic acid dehydrogenase linked to an γA immunoglobulin. Experientia *23:* 593 (1967).

3 Kindmark, C.O.: Atypical lactate dehydrogenase isoenzyme pattern caused by immunoglobulin G interaction. Scan. J. clin. Lab. Invest. *24:* 49–53 (1969).

4 Biewenga, J.; Thys, L.G.: Lactate dehydrogenase isoenzyme(s) linked to IgA immunoglobulin in a patient with a myocardial infarction. Clinica chim. Acta *27:* 293–299 (1970).

5 Biewenga, J.: Serum lactate dehydrogenase isoenzymes linked to immunoglobulin A. Clinica chim. Acta *40:* 407–411 (1972).

6 Nagamine, M.: Lactate dehydrogenase isoenzymes linked to immunoglobulin A in two cases. Clinica chim. Acta *36:* 139–144 (1972).

7 Biewenga, J.: Thesis, University of Amsterdam, The Netherlands (1974).

8 Markel, S.O.; Janick, S.L.: Complexing of lactate dehydrogenase isoenzymes with immunoglobulin A of the kappa class. Am. J. clin. Path. *61:* 328–332 (1974).

9 Thomas, D.W.; Rosen, S.W.; Kahn, C.R.: Macromolecular lactic acid dehydrogenase. Ann. intern. Med. *81:* 434–439 (1974).

10 Biewenga, J.; Feltkamp, T.E.W.: LDH-IgA immunoglobulin complexes in human serum. Clinica chim. Acta *58:* 239–249 (1975).

11 Biewenga, J.; Feltkamp, T.E.W.: Lactate dehydrogenase (LDH)-IgG_3 immunoglobulin complexes in human serum. Clinica chim. Acta *64:* 101–116 (1975).

12 Biewenga, J.: Heterogeneity in electrophoretic mobility and molecular size of lactate dehydrogenase-immunoglobulin G complexes. Clinica chim. Acta *76:* 149–153 (1977).

13 Trocha, P.J.: Lactate dehydrogenase isoenzymes linked to beta-lipoproteins and immunoglobulin A. Clin. Chem. *23:* 1780–1783 (1977).

14 Biewenga, J.; Loghem, E. van: Antigenic analysis of the IgA component of LDH-IgA immunoglobulin complexes. Clinica chim. Acta *82:* 201–204 (1978).

15 Leroux-Roels, G.G.; Wieme, R.J.; Broe, M.E. de: Occurrence of enzyme-immunoglobulin complexes in chronic inflammatory bowel disease. J. Lab. clin. Med. *97:* 316–321 (1981).

16 Peters, O.; Gorus, F.; Camp, B. van: NAD^+-dissociable macromolecular lactate dehydrogenase. Clin. Chem. *28:* 1826 (1982).

17 Gorus, F.; Aelbrecht, W.; Camp, B. van: Circulating IgG-LD complex, dissociable by addition of NAD^+. Clin. Chem. *28:* 236–239 (1982).

18 Weijers, R.N.M.; Mulder, J.; Kruijswijk, H.: Partial characterization, properties, and clinical significance of a lactate dehydrogenase-immunoglobulin A_κ complex in serum. Clin. Chem. *29:* 272–278 (1983).

19 Fujita, K.; Takeya, C.; Saito, T.; Sakurabayashi, I.: Macro lactate dehydrogenase. An LDH-immunoglobulin M complex that inhibits lactate dehydrogenase activity in a patient's serum. Clinica chim. Acta *140:* 183–195 (1984).

20 Wickus, G.G.; Smith, M.J.: Rapid loss of lactate dehydrogenase isoenzyme activity in serum by cold-induced formation of immunoglobulin G-lactate dehydrogenase complex. Clin. Chem. *30:* 11–17 (1984).

21 Podlasek, S.J.; McPherson, R.A.; Threatte, G.A.: Specificity of auto-antibodies to lactate dehydrogenase isoenzyme subunits. Clin. chem. *31:* 527–532 (1985).

22 Jockers-Wretou, E.; Plessing, E.: Atypical serum creatine kinase isoenzyme pattern caused by complexing of creatine kinase-BB with immunoglobulins G and A. J. clin. Chem. clin. Biochem. *17:* 731–737 (1979).

23 Bohner, J.; Stein, W.; Kuhlmann, E.; Eggstein, M.: Serum creatine kinase BB linked to immunoglobulin G. Clinica chim. Acta *97:* 83–88 (1979).

24 Stein, W.; Bohner, J.; Eggstein, M.: Eigenschaften und Bedeutung von Macro-Kreatinkinasen. J. clin. Chem. clin. Biochem. *17:* 739–745 (1979).

25 Stein, W.; Bohner, J.; Krais, J.; Müller, M.; Steinhart, R.; Eggstein, M.: Macro creatine kinase BB. Evidence for specific binding between creatine kinase BB and immunoglobulin G. J. clin. Chem. clin. Biochem. *19:* 925–930 (1981).

26 Bohner, J.; Stein, W.; Renn, W.; Steinhart, R.; Eggstein, M.: Stability of macro creatine kinases and creatine kinase isoenzymes compared. Heat inactivation test for determination of thermostable creatine kinases. J. clin. Chem. clin. Biochem. *19:* 1021–1026 (1981).

27 Bohner, J.; Stein, W.; Steinhart, R.; Würzburg, U.; Eggstein, M.: Macro creatine kinases. Results of isoenzyme electrophoresis and differentiation of the immunoglobulin-bound type by radioassay. Clin. Chem. *28:* 618–623 (1982).
28 Urdal, P.: Macro creatine kinase BB isoenzyme in serum: most likely an antigen-autoantibody complex. Scan. J. clin. Invest. *41:* 499–505 (1981).
29 Lang, H.; Würzburg, U.: Creatine kinase, an enzyme of many forms (Review). Clin. Chem. *28:* 1439–1447 (1982).
30 Wu, A.H.B.; Herson, V.C.; Bowers, G.N., Jr.: Macro creatine kinase types 1 and 2. Clinical significance in neonates and children as compared with adults. Clin. Chem. *29:* 201–204 (1983).

R.N.M. Weijers, PhD, Department of Clinical Chemistry,
Onze Lieve Vrouwe Gasthuis, 1e Oosterparkstraat 179,
NL–1091 HA Amsterdam (The Netherlands)

Adv. clin. Enzymol., vol. 3, pp. 168–176 (Karger, Basel 1986)

The Role of Phagocyte Proteinases in the Pathobiochemistry of Inflammation

Hermann Lang[a], *Hans Fritz*[b]

[a] Biochemical Research Institute, E. Merck, Darmstadt, FRG;
[b] Institute of Clinical Biochemistry, University of Munich, FRG

The basic mechanism of the inflammatory process – as seen by the biochemist – is summarized in figure 1. A primary stimulus, which can be an injury or an infection (top), triggers the synthesis and release of many mediators (centre). These then activate the different systems of inflammation of either humoral or cellular nature. The sum of all reactions exerted by these systems is the so-called 'inflammatory response' (periphery). The following discussion concentrates on the pathobiochemical role of the phagocytes and their proteinases in the inflammatory reaction. The phagocytes concerned are polymorphonuclear granulocytes or neutrophils, macrophages, mast cells, fibroblasts, and some types of endothelial cells.

The mechanisms leading to the accumulation of mobile phagocytes in the focus of inflammation, and their concomitant activation, are shown in figure 2. The mediators attracting the mobile phagocytes are called chemotaxins. Powerful chemotaxins are (besides endotoxins) factors from the complement cascade (especially the 'anaphylatoxins' C 3a and C 5a), plasma kallikrein, factors from the clotting casade (fibrin monomers), and from the fibrinolytic system (fibrin and fibrinogen split products). The movement of phagocytes is supported by the increased vascular permeability induced by different kinins (e.g. bradykinin).

The view again is focused in figure 3 onto a single phagocyte – here a neutrophil. At the focus of inflammation, after stimulation by the mediators, it exerts 3 types of activity: first it produces more mediators, which cross-activate other phagocytes (the neutrophil, for example, produces macrophage activating factor; MAF). Secondly it produces, in a process called

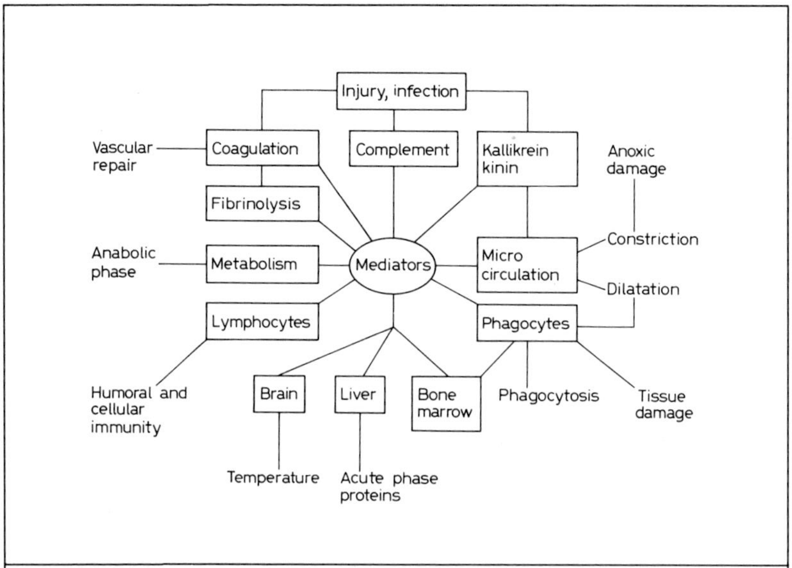

1

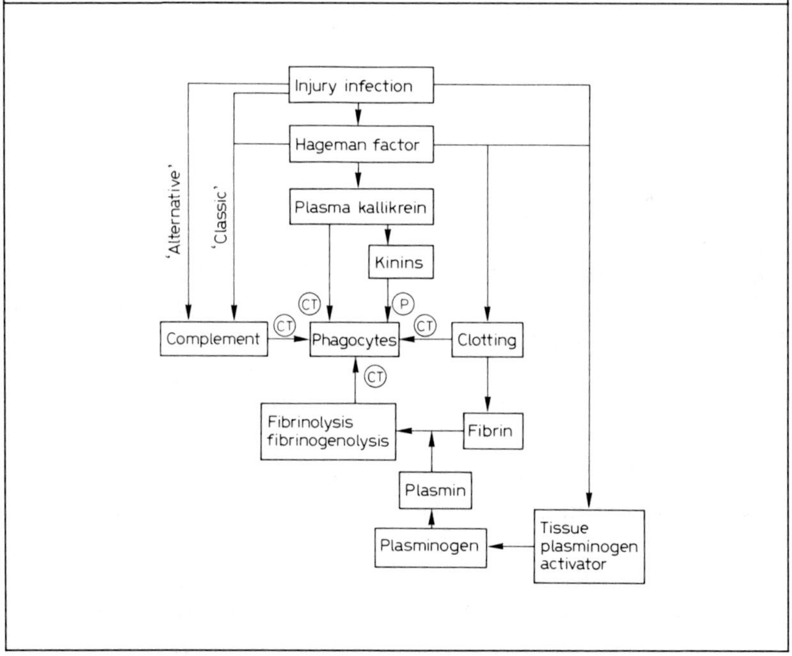

2

Fig. 1. Inflammation – simplified.

Fig. 2. The inflammatory response: accumulation and activation of phagocytes. CT = Chemotaxins and activators, P = increased vascular permeability.

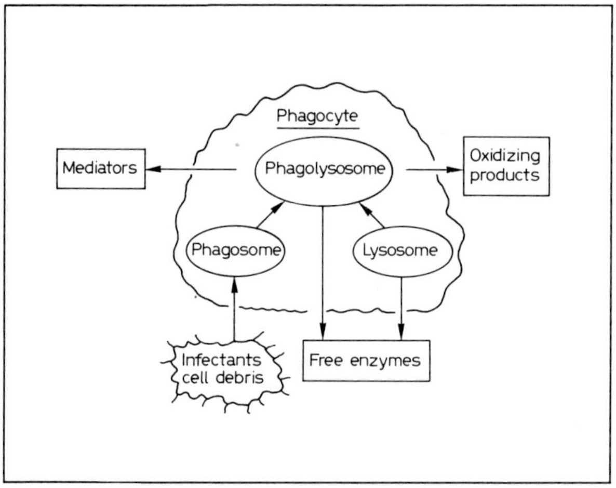

Fig. 3. The inflammatory response: action of stimulated phagocytes.

'metabolic burst', oxidizing agents (oxygen and hydroxyl radicals) and releases oxidizing enzymes (myeloperoxidase). Thirdly, it begins phagocytosis, internalization and degradation of infectious agents plus cell debris. These materials are engulfed into the cell in vesicles called phagosomes. The phagosomes join with lysosomes to form phagolysosomes, in which the enzymatic degradation takes place. During the phagocytic process some of the lysosomal enzymes (proteases, glycosidases, etc.) leak to the outside of the cell: the more intensive the stimulation, the more enzyme discharged. Following excessive stimulation the neutrophil may discharge all of its lysosomal enzymes – a process called 'degranulation' – or it may even disintegrate totally.

Figure 4 shows the extracellular action of proteinases discharged by a phagocyte. The free, extracellular proteinases present a highly dangerous potential, since many of the neutral proteinases are able to cleave proteins in their native state. This proteolytic action leads to tissue destruction, breakdown of regulatory proteins (e.g. consumption of antithrombin III), and to the liberation of more mediators, which in turn amplify the destructive process, which in the worst case can lead to fatal consequences, e.g. in posttraumatic lung shock. On the other hand, the activity of extracellular lysosomal proteinases is checked by proteinase inhibitors. These inhibitors

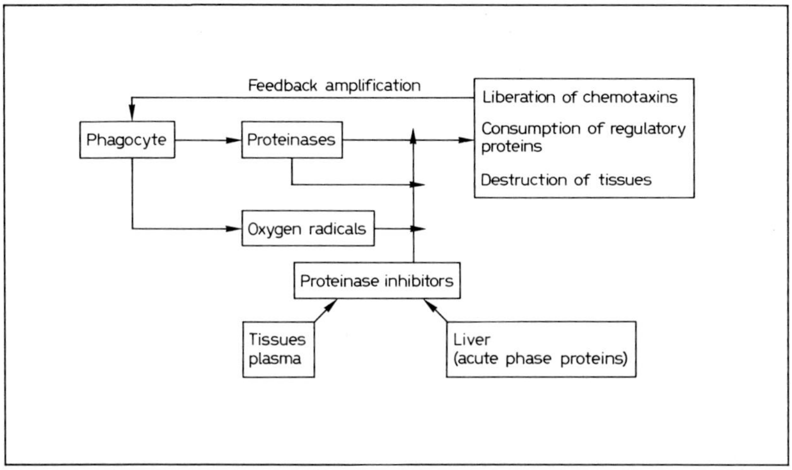

Fig. 4. The inflammatory response: extracellular action of phagocyte proteinases and radicals.

primarily diffuse from surrounding tissues and blood into the focus of inflammation, in a secondary reaction inhibitors are produced by the liver as acute phase proteins. To complicate the picture, the inhibitors in turn can be inactivated by the proteinases and the oxidizing agents from the phagocyte. The balance between the concentration of free, active enzyme and free, active inhibitor determines whether the deleterious actions of extracellular proteolysis will take place.

The Proteinases from Phagocytes and Their Inhibitors

A summary of the phagocyte proteinases is given in table I. The enzymes are characterized according to the producing cell type, with respect to their function (lysosomal or secretory) and to their activity range (neutral or acid). The most potent and most abundant proteinase is the elastase from neutrophils (PMN elastase). The enzyme is a highly unspecific, neutral proteinase. It is able, therefore, to cleave a broad range of structural and functional proteins in their native state. Neutrophils contain 3 μg of PMN elastase per million cells: the daily production of 10^{11} neutrophils of a healthy subject, therefore, contains 300 mg of elastase.

Table I. The proteinases of phagocytes

Phagocyte	Proteinase	Origin	Activity
Neutrophil	PMN elastase	L	N
	collagenase	L	N
	cathepsin G	L	N
	cysteine proteinases	L	A
Macrophage	MAC elastase	L	N
	cysteine proteinases	L	A
	tissue plasminogen activator	S	N
Mast cell	chymase	?	N
	tryptase	?	N
	cysteine proteinases	L	A
Endothelial cell	cysteine proteinases	L	A
	tissue plasminogen activator	S	N

L = Lysosomal; S = secrectory; N = neutral; A = acid pH.

Table II. Elastin degrading enzymes: $M_r \sim 25,000–30,000$

Origin	Proteinase type	Proenzymes	Multiple forms	Cleavage preference	Inhibitors
Neutrophil elastase	serine, endo	no	4	Val-X Met-X	α_1-PI α_2-M
Macrophage elastase	metallo, Zn^{++}	no	?	H-X	EDTA
Pancreatic elastase	serine, endo	yes	2	Ala-X	α_1-PI α_2-M
Bacterial elastases	serine, endo	yes/no	yes	H-X	

H = Hydrophilic amino acid residue; α_1-PI = alpha-1-proteinase inhibitor; α_2-M = alpha-2-macroglobulin.

Table III. Proteinase inhibitors in human blood [modified from Travis and Salvesen, 1983]

Inhibitor	Primary target enzymes	Acute phase protein
α_1-Proteinase inhibitor[1]	PMN-elastase, chymotrypsin	yes
α_2-Macroglobulin	trypsin, chymotrypsin, elastase, plasmin, plasma kallikrein	no
α_1-Antichymotrypsin	chymotrypsin, cathepsin G	yes
β_1-Collagenase inhibitor	collagenase	?
Inter-α-trypsin inhibitor	trypsin	no
Antithrombin III	thrombin, factor Xa	no
α_2-Antiplasmin	plasmin	yes
C1-Inactivator	complement esterase C1, plasma kallikrein, Hageman factor fragments	yes

[1] Ex 'α_1-antitrypsin'.

The different types of enzymes termed 'elastase' are characterized in table II. These enzymes definitely are distinct proteinases; they can easily be differentiated by biochemical and immunological methods. The terminology is misleading, the common name 'elastase' of course stems from the time of the discovery of these enzymes, when they were found to exert elastin-degrading activity.

The human organism provides a broad spectrum of inhibitors to control the potentially dangerous activity of the phagocyte (and other) proteinases. Table III summarizes the proteinase inhibitors present in human blood. At the top of the list are the inhibitors directed against the lysosomal proteinases. The most abundant inhibitor is α_1-proteinase inhibitor, formerly called α_1-antitrypsin. Meanwhile, it has been established that it is the physiological antagonist of PMN elastase. It is an acute-phase protein, and is produced by the liver in very large quantities following an inflammatory stimulus. The lower part of the table contains the inhibitors directed against the proteinases of the clotting, fibrinolytic, and complement cascades.

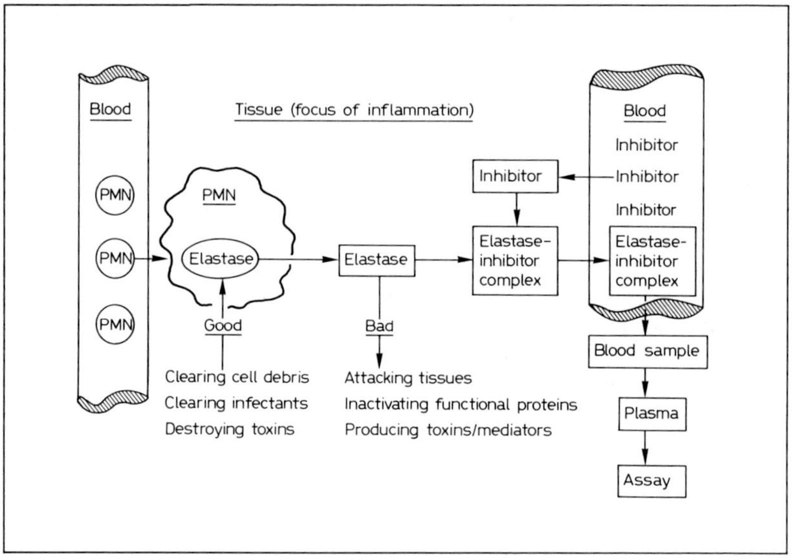

Fig. 5. Plasma elastase-inhibitor complex as marker for inflammatory process in tissue.

The Value of PMN Elastase as a Biochemical 'Inflammation Marker'

The working concept of our inflammation marker project is given in figure 5: neutrophils are attracted into the focus of inflammation from the blood, and upon stimulation they begin phagocytic activity. During this process, elastase is discharged to the surrounding tissue. When the concentration of elastase released is higher than the local concentration of α_1-proteinase inhibitor (and other inhibitors), the deleterious action of free elastase will take place. The amount of damage is related to the intensity of neutrophil stimulation. With increasing inhibitor concentration, finally all elastase discharged is bound by the inhibitor, and the inactive inhibitor complex diffuses into the blood. It is cleared from the blood by the reticulo-endothelial system with a half-life of 1 h.

We have developed a specific sandwich immunoassay for the quantitative measurement of elastase-α_1-proteinase inhibitor complex in plasma. Our concept is that the quantitation of the elastase inhibitor complex in plasma gives a direct measure of the neutrophil activity in the focus of

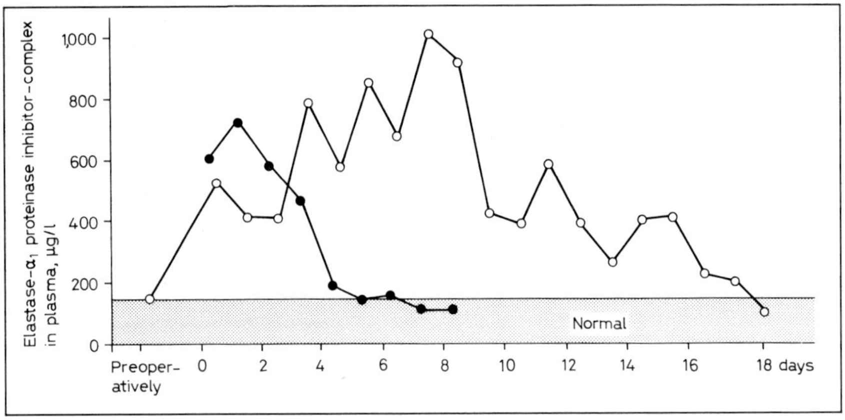

Fig. 6. Neutrophil activity in post-operative inflammation. [From Inthorn and Jochum, personal communication 1985].

inflammation: in other words, the concentration of the elastase inhibitor complex in plasma is a marker for the activity of the inflammatory process in tissues. Two pieces of evidence are given concerning the validity of this concept.

One example of the diagnostic value of measuring elastase inhibitor complex in the plasma is given in figure 6. The elastase inhibitor concentration was measured in 2 patients undergoing gastrectomy because of gastric cancer. In both patients the operative trauma caused a peak on the first and second postoperative days, respectively. In the patient without complications (dots) the concentration decreased within 5 days to normal. The other patient (open circles) developed insufficiency of the surgical suture resulting in severe postoperative inflammation. The elastase-inhibitor concentration started to increase on the fourth postoperative day, the clinical signs of inflammation reached a maximum on the ninth postoperative day. After re-operation (day 12) the patient recovered with concomitant decrease to normal levels. We take this as definite evidence that the plasma elastase-inhibitor complex is indeed a valid marker for the activity of inflammatory processes.

Another piece of evidence is the experimental proof that the action of free elastase is one of the pathogenetic mechanisms leading to the inflammatory response. Table IV gives data from experimental septicaemia in pigs. The measure was the survival time of the animals, which was followed

Table IV. The inflammatory response: antagonistic action of proteinase inhibitors; survival time in experimental septicaemia in pigs; inhibitor: 'Eglin' (leech inhibitor, cloned in *E. coli*) [modified from Jochum et al., 1985]

Group No.	n	*E. coli* dose $\times 10^9$ cells	Inhibitor dose mg/kg $\times$ h	Mean survival time
1	10	–	–	
2	8	3	–	
3	9	30	–	
4	5	–	3.85	
5	5	3	3.85	
7	6	30	3.85	

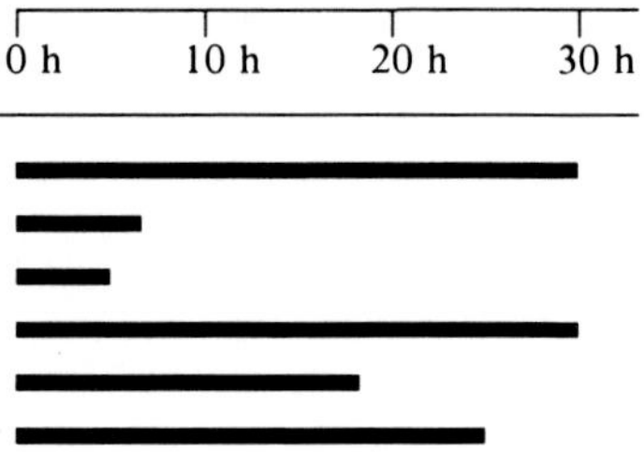

up to 30 h. The septicaemic animals were injected with *Escherichia coli,* which produced a catastrophic inflammatory response: the mean survival time was dose related and was reduced to 7 and 5 h, respectively (groups 2 and 3). When septicaemic animals were given a potent elastase inhibitor together with *E. coli,* the mean survival time was increased to 19 and 25 h, respectively (groups 5 and 7). The inhibitor used was Eglin, a protein from the leech with a very high affinity for PMN elastase. We take these results as evidence that elastase is indeed a pathogenetic principle in producing the inflammatory response.

References

Jochum, M.; Welter, H.F.; Wiesinger, H.; Siebeck, M.; Thetter, O.; Fritz, H.: Preliminary experiments with an inhibitor of lysosomal elastase and cathepsin G ('Eglin' from the leech) during septicemia in pigs; in Stelzner, Chirurgisches Forum 1985, p. 43 (Springer, Berlin 1985).

Travis, J.; Salvesen, G.S.: Human plasma proteinase inhibitors. Ann. Rev. Biochem. *52:* 655–709 (1983).

Hermann Lang, PhD, Biochemical Research Institute, E. Merck,
PO Box 41 19, D-6100 Darmstadt 1 (FRG)

Adv. clin. Enzymol., vol. 3, pp. 177–183 (Karger, Basel 1986)

Enzymes and Isoenzymes Involved in the Release of Vasoactive Peptides

Marc van Sande, Simon Scharpé, Dirk Hendriks, Koenraad Van Camp
Faculty of Medicine, University of Antwerp, Wilrijk, Belgium

The enzymes and isoenzymes involved in the release of vasoactive peptides are receiving a sustained and rapidly growing interest in biomedical research and have been the subject of excellent reviews [1–11]. In this paper we focus special attention on important recent findings on renin, angiotensin-converting enzyme, kallikrein, carboxypeptidase N and their different molecular forms. A schematic outline of some of their functional relationships is given in figure 1.

Renin (EC 3.4.99.19) is an aspartyl protease. It differs from other aspartyl proteases, such as pepsin, in that its activity is not optimal at acid pH. Renin also acts on a much more restricted set of substrates than other aspartyl proteases. The human gene encoding renin has been isolated and characterized [12]. The complete amino acid sequence of renin has been established, and the results indicate a close similarity of renin with acid proteases in its active site structure [13]. In man, renin is present in different forms: pre-prorenin, prorenin, big renin, inactive renin and active renin. It seems more interesting to known that renin, or more strictly, renin-like enzymes have been detected in other tissues than the kidney. Aorta, liver, uterus and even human brain have been studied for these enzymes [14].

Plasma renin can be separated by isoelectric focusing in 6 different forms. The relative properties of the renin forms are influenced by agonists, altered salt diet, acute and chronic renal artery constriction. Furthermore, evidence has been presented that the multiple renin forms are degraded by the liver at different rates. Whether these 6 distinct forms of renin are isoenzymes or are due to variations in glycosylation is undetermined [15].

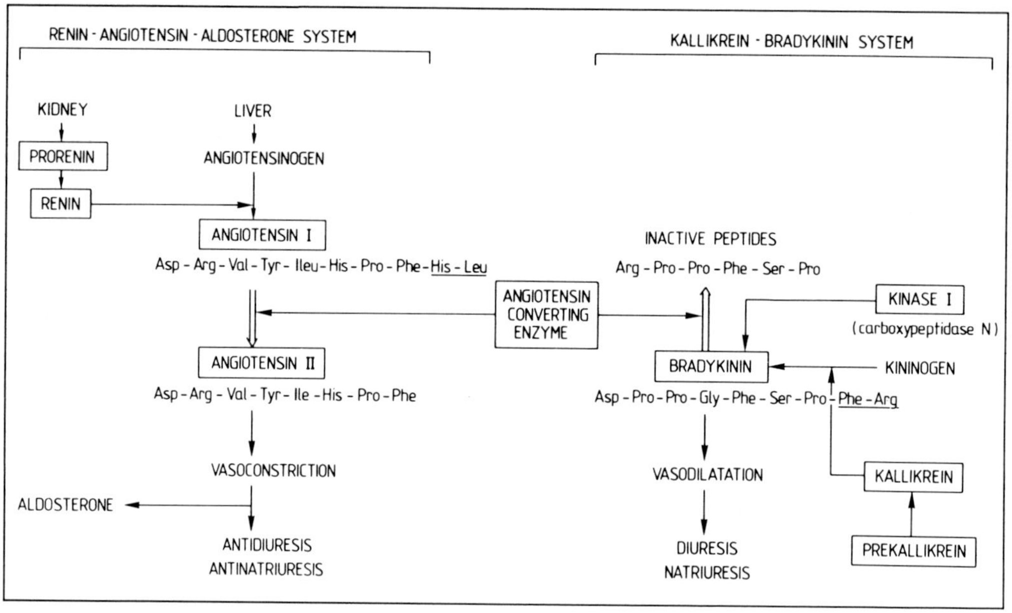

Fig. 1. Schematic representation of the renin-angiotensin and the kallikrein-kinin systems.

Specific inhibitors of renin are likely to play a significant role in the therapy of hypertension. The most promising inhibitors of renin include several classes of compounds: (a) specific antibodies; (b) general peptide inhibitors of acid proteases, and (c) angiotensinogen analogs and peptides that are related to the terminal amino acid sequence of renin's precursor, prorenin [16]. Although these ideas are attractive, until now no really practical solution for antihypertensive therapy in man has been found. Very recently, it has been shown that the increase of renin activity in plasma of patients with renal failure may be related to the absence of a normally-occurring renin-inhibiting lipid component of plasma [17]. The composition of the renin-inhibiting component is still unknown.

The next enzyme we discuss is angiotensin-I-converting enzyme (EC 3.4.15.1; ACE, kininase II).

In the early 1960s, ACE was established as a dipeptidyl carboxypeptidase, which cleaves both angiotensin I and bradykinin. The name dipeptidyl carboxypeptidase is not entirely correct. The Enzyme Commission named the enzyme dipeptide hydrolase or peptidyl dipeptidase. Very many publications deal with ACE. A synopsis of all available information: (a) the formation of the vasoconstrictive angiotensin II; (b) the value of the serum ACE determination in active sarcoidosis, and (c) the finding of ACE inhibitors, would exceed the scope of this paper. Angiotensin II has a variety of actions, other than vasoconstriction, e.g. stimulation of aldosterone, antidiuretic hormone, thirst, inhibition of renin release by the kidney and probably stimulation of ACTH release. Furthermore, strong evidence exists for the presence of a renin-angiotensin system in brain, independent of the peripheral one. The enzyme presumably recognizes other peptides as substrate, including enkephalin, met-enkephalin, met-enkephalin-Arg^6-Phe^7, and substance P. In some cases, ACE acts as a tripeptidyl carboxypeptidase [18], an observation that complicates the original nomenclature given to the enzyme and in fact underscores our limited knowledge of this enzyme. Homogeneous ACE hydrolyses the NH_2- and COOH-terminal tripeptides from the luteinizing hormone-releasing hormone [19] and it cleaves also the COOH-terminal tripeptides from substance P [20]. Very recently, it has been shown that ACE cleaves des-Arg^9-bradykinin, and, although to a lesser extent, angiotensin II [21]. It should be remarked that ACE acts in these cases as a tripeptidyl carboxypeptidase.

It was speculated that by inhibiting ACE (captopril, enalapril, etc.), bradykinin, as vasodilator, could play an important role in hypertension. However, it has been shown now that bradykinin can be cleaved by other enzymes, such as the recently isolated neutral metalloendopeptidase (EC 3.4.24.11; neutral endopeptidase, NEP, enkephalinase) which was initially assumed to be, like ACE, a peptidyl dipeptidase [22].

We studied the electrophoretic behaviour of ACE in crude tissue homogenates [23, 24], using agar and polyacrylamide gel. Results obtained showed that benign prostatic hyperplasia has the highest ACE activity of all the samples studied. Seminal plasma gives two important peaks, one with the same relative mobility of prostate, the other coinciding with the ACE of seminal vesicles. The differences in relative mobility of ACE are a function of the number of sialic acid groups in the ACE molecule [unpubl. results]. We found a shift of the electrophoretic mobilities after treatment with neuraminidase (fig. 2). The heterogeneity of purified ACE has been reported by several authors, using different separating techniques.

Recently, it has been shown that testicular ACE possesses distinct physicochemical and immunological properties and a susceptibility to hormonal regulation [25, 26]. The function of ACE in the male reproductive system remains unknown.

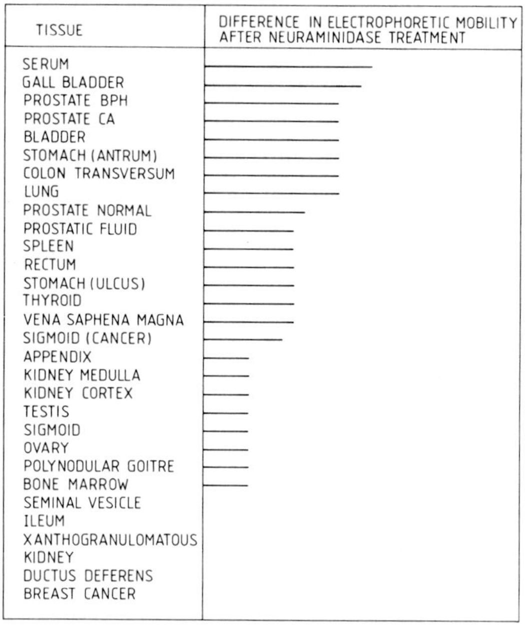

Fig. 2. Influence of neuraminidase treatment on the electrophoretic behaviour in agar gel of angiotensin-converting enzyme from human tissues. The solid lines represent, in arbitrary units, the distance between the major ACE fraction before and after neuraminidase treatment.

Kallikrein (EC 3.4.21.8) denotes those serine proteases which liberate kinins from kininogen by limited proteolysis and which have little or no proteolytic activity on other proteins. All mammals so far studied have two kallikrein types: plasma kallikrein, and glandular kallikrein, which is found in exocrine glands and their secretions. Plasma and glandular kallikreins differ in (a) physicochemical properties; (b) rates of reaction with kininogens and synthetic substrates; (c) kinins produced, and (d) response to a variety of synthetic and natural inhibitors.

Known functions of kallikrein include: (a) generation of bradykinin from kininogens, (b) activation of factor XII, and (c) a chemotactic activity towards neutrophil leucocytes.

Kallikrein circulates in the blood as a precursor, prekallikrein, which is hydrolysed by plasma proteases (proteolytic fragments derived from factor XIIa). Purified prekallikrein from human plasma gives 7 peaks by means of isoelectric focusing. The heterogeneity may be caused partially or wholly by differences in the sialic acid content of the isoelectric forms [27]. Multiple forms of glandular and human urinary kallikreins have been obtained after isoelectric focusing [28].

Another enzyme that can hydrolyse bradykinin has been described: carboxypeptidase N [29]. Although first described in 1962, it is only recently that full attention has been given to this enzyme. Carboxypeptidase N (EC 3.4.17.3) is a plasma enzyme with a molecular weight of about 300,000 and with a concentration in human plasma of about 10^{-7} mol/l. The name kininase I is also used because of its function in the metabolism of kinins. It is a metallopeptidase, migrating as an α_1-globulin. It is composed of heavy and light chains. The last have no carbohydrate moiety. After isoelectrofocusing of carboxypeptidase N from human serum, two forms of the enzyme are detected. The enzyme inactivates physiologically active peptides, such as bradykinin and the anaphylatoxins C3a and C5a, by releasing the COOH-terminal arginine. For this reason the Enzyme Commission called it arginine carboxypeptidase. However, it cleaves the C-terminal lysine from fibrin degradation products FDP-6A and 6D. A potent inhibitor of the enzyme is *DL*-2-mercapto-methyl-3-guanidinoethyl propionic acid. Carboxypeptidase N appears to be a metabolic crossroad for all peptides generated by trypsin-like enzymes and terminated by a basic amino acid. Being a plasma protein, carboxypeptidase N is available in the peripheral blood and eventually crosses the capillary wall under the action of pro-inflammatory peptides which increase capillary permeability. Once in the exudate, carboxypeptidase N may take part in the metabolism of these pro-inflammatory peptides and thus in the control of their biological activities.

References

1 Roche, E.; Silva, M.: Biochemistry and pharmacology of bradykinin; in Schachter, Polypeptides which affect smooth muslces and blood vessels, pp. 210–231 (Pergamon Press, Oxford 1960).

2 Schachter, M.: Kinins – a group of active peptides. A. Rev. Pharmac. *4:* 281–292 (1964).

3 Frey, E.K.; Kraut, H.; Werle, E.: Das Kallikrein Kinin System und seine Inhibitoren (Enke, Stuttgart 1968).

4 Pisano, J.J.: Chemistry and biology of the kallikrein-kinin system; in Reich, Rifkin, Shaw, Proteases and biological control, pp. 199–222 (Cold Spring Harbor Laboratory, Cold Spring Harbor 1975).

5 Peach, M.J.: Renin-angiotensin system. Biochemistry and mechanisms of action. Physiol. Rev. *57:* 313–370 (1977).

6 Levinski, N.G.: The renal kallikrein-kinin system. Circulation Res. *44:* 441–451 (1979).

7 Soffer, R.L.: Angiotensin-converting enzyme and the regulation of vasoactive peptides. A. Rev. Biochem. *45:* 73–94 (1976).

8 Erdös, E.G.: Bradykinin, kallidin and kallikrein. Handb. exp. Pharm. *25:* suppl., (Springer, Berlin 1979).

9 Regoli, D.; Barabé, J.: Pharmacology of bradykinin and related kinins. Pharmac. Rev. *32:* 1–46 (1980).

10 Schachter, M.: Kallikreins (kininogenases). A group of serine proteases with bioregulatory actions. Pharmac. Rev. *31:* 1–17 (1980).

11 Gross, E.; Vogel, V.: Enzymatic release of vasoactive peptides (Raven Press, New York 1980).

12 Hardman, J.A.; Hort, Y.J.; Catanzaro, D.F.; Tellam, J.T.; Baxter, J.D.; Morris, B.; Shine, J.: Primary structure of the human renin gene. DNA *3:* 457–468 (1984).

13 Inagami, T.; Misoni, K.S.; Chang, J.-J.C.; Takii, Y.: Biochemistry of renin. Biochem. Soc. Trans. *12:* 951–953 (1984).

14 Re, R.; Bryan, S.E.: Functional intracellular renin-angiotensin systems may exist in multiple tissues. Clin. Exp. Theory Pract. *6A:* 1739–1742 (1984).

15 Malvin, R.Z.; Katz, S.A.; Shier, D.N.; Sessler, F.M.: Functional significances of multiple forms of renin. Clin. Exp. Theory Pract. *6A:* 1781–1786 (1984).

16 Zusman, R.M.; Burton, J.; Christensen, D.; Nussberger, J.; Dodds, A.; Haber, E.: Hemodynamic effects of a competitve renin inhibitory peptide in humans. Evidence of multiple mechanisms of action. Trans. Am. Ass. Physns. *96:* 365–374 (1983).

17 Kotchen, T.A.; Talwalkar, R.T.; Kaul, K.: Identification of renin inhibitors in normal and uremic plasma. J. Lab. clin. Med. *105:* 286–293 (1985).

18 Inokuchi, J.I.; Nagamatsu, A.: Tripeptidyl carboxypeptidase activity of kininase II (angiotensin-converting enzyme). Biochim. biophys. Acta *662:* 300–307 (1981).

19 Skidgel, R.A.; Erdös. E.G.: Novel activity of human angiotensin I converting enzyme. Release of the NH_2- and COOH-terminal tripeptides from the luteinizing hormone-releasing hormone. Proc. natn. Acad. Sci. USA *82:* 1025–1029 (1985).

20 Pernow, B.: Studies on substance P. Purification, occurence and biological actions. Acta physiol. scand. *29:* suppl. 105, pp. 1–90 (1953).

21 Oshima, G.; Hiraga, Y.; Shirono, K.; Ohishi, S.; Sakakibara, S.; Kinoshita, T.: Cleavage of des-Arg^9-bradykinin by angiotensin-I-converting enzyme from pig kidney cortex. Experientia *41:* 325–328 (1985).

22 Erdös, E.G.; Schultz, W.W.; Gafford, J.T.; Defendini, R.: Neutral metalloendopeptidase in human male genital tract. Comparison to angiotensin-I-converting enzyme. Lab. Invest. *52:* 437–447 (1985).

23 Sande, M. van; Scharpé, S.L.; Neels, H.M.; Van Camp, K.O.: Distribution of angiotensin converting enzyme in human tissues. Clinica chim. Acta *147:* 255–260 (1985).

24 Sande, M. van; Scharpé, S.L.; Neels, H.M.: Multiple forms of angiotensin-converting enzyme in human tissues and fluids. J. clin. Chem. clin. Biochem. *23:* 381–386 (1985).

25 Velletri, D.P.: Testicular angiotensin-I-converting enzyme (EC 3.4.15.1). Life Sci. *36:* 1597–1608 (1985).

26 Lanzillo, J.L.; Dasarathy, Y.; Stevens, J.; Wayne Bardin, C.; Fanburg, B.L.: Human testicular angiotensin-converting enzyme is a mixture of two molecular forms. Only one is similar to the seminal plasma enzyme. Biochem. biophys. Res. Commun. *128:* 457–463 (1985).
27 Hojima, Y.; Pierce, J.V.; Pisano, J.J.: Purification and characterization of multiple forms of human plasma prekallikrein. J. biol. Chem. *260:* 400–406 (1985).
28 Hojima, Y.; Matsuda, Y.; Moriwaki, C.; Moriva, H.: Homogeneity and multiplicity forms of glandular kallikrein. Chem. Pharm. Bull. *23:* 1120–1127 (1975).
29 Erdös, E.G.; Renefren, A.G.; Sloane, E.M.; Wohler, J.R.: Enzymatic studies on bradykinin and similar peptides. Ann. N.Y. Acad. Sci. *104:* 222–234 (1963).

S.L. Scharpé, MD, Unit of Clinical Biochemistry, Faculty of Medicine, University of Antwerp, Universiteitsplein 1, B–2610 Wilrijk (Belgium)

Adv. clin. Enzymol., vol. 3, pp. 184–201 (Karger, Basel 1986)

Lysosomal Enzyme Replacement Therapy by Displacement Bone Marrow Transplantation with Immunoprophylaxis

J.R. Hobbs, K. Hugh-Jones, J.D. Chambers, S. Williamson, C.J.C. Downie, P. Shaw, M.L. Knapp, R. Knight, T.R. Rogers

Westminster Bone Marrow Team, Westminster Hospital, London, UK

Earlier experience in bone marrow transplantation [1–3] enabled observations of the transfer of normal components from donor to genetically deficient cells, both in vitro and in vivo, leading to the concept that displacement bone marrow transplantation (DBMT) might help patients with certain genetic diseases [4, 5]. This concept is summarized in table I and figure 1, and it has been predicted that some 95 diseases from among some 1,300 where the metabolic defect is reasonably understood, might respond to such treatment [4–6]; already, attempts have been made in 48 such diseases and on current evidence it would seem that 39 of those are correctable (table II), excluding the mucopolysaccharidoses. For reasons reviewed elsewhere [6], only the displacement method can be recommended for 35 of those diseases and, indeed, when the donor is other than a matched family member (sharing the same genetic haplotypes for tissue types), DBMT is recommended in all cases where it is wished to install an enzyme factory of donor type within the bone marrow of the recipient.

Of the diseases listed in table II numbers 22, 28, 35, 36, 37 and 38 are known to arise from impairment of enzyme activity within the lysosome, where correction is achievable, and also in the mucopolysaccharidoses, 29–34, where it is clear that the systemic disease in certain sites is fully correctable, but in others such as the nervous system it is still too early to know whether there is enough correction to ensure a reasonable lifestyle for the patient.

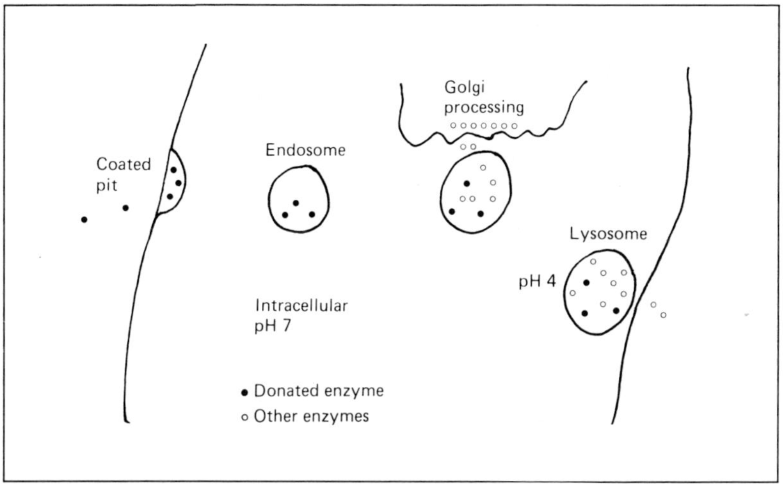

Fig. 1. Enzyme donation at the cellular level. Coated pits carry receptors for lysosomal enzymes that become intracellular within endosomes. Further enzymes can be added in the Golgi region to form the mature lysosome wherein the reactions can occur at pH 4. Other proteins and enzymes can gain the cytosol in a similar manner.

Table I. Rationale of displacement bone marrow transplantation

1	The error is expressed by bone marrow stem cells
2	Displace abnormal marrow by DONOR marrow
3	Donor marrow synthesises missing component
4	Recipient should be tolerized to that component
5	Daily leucocyte production 50–300 g
6	Leucocytes circulate, most spill component in tissue spaces, some may deliver cell-to-cell
7	Component can be pinocytosed by defective tissues
8	Component may be able to 'work' after pinocytosis, etc.

It should be emphasized that because the engrafted bone marrow generates immunologically competent cells, these can sometimes attack the new host to cause graft-vs.-host disease (GVHD) and this can result in the death or severe handicap of the recipient. For such reasons, BMT has always been reserved for otherwise fatal disease for which there are no successful alternative treatments. Guidance as to the optimal time for

Table II. Inborn errors of metabolism successfully corrected by bone marrow transplantation (December 1984)

		n[1]
	Of lymphocyte function	
1	Reticular dysgenesis	1
2	Severe combined immunodeficiency (sex-linked, helper-T)	60
3	'Swiss-type' SCID (autosomal recessive, T and B)	14
4	Severe combined immunodeficiency with cartilage-hair dysplasia	2
5	Late adenosine deaminase deficiency	12
6	Late purine nucleoside phosphorylase deficiency	4
7	Late DiGeorge syndrome	2
8	Connatal GVHD	5
9	Wiskott-Aldrich syndrome	11
10	Helper T cell deficiency (autosomal recessive)	1
11	Bare lymphocyte syndrome (without DR expression)	7
12	Bare lymphocyte syndrome (without HLA-A,B expression)	5
13	Bare lymphocyte syndrome (without class I or II)	0
14	Severe lymphokine deficiency (i) chronic mucocutaneous candidiasis	2
15	Severe lymphokine deficiency (ii) others (interferon)	2
16	Late onset T cell deficiency; Nezelof/Matsaniotis	5
17	Late T cell defects	2
18	Interleukin receptor deficiency	1
19	Biotinidase deficiency	2
20	Adult combined immunodeficiency	1
	Of phagocytosis	
21	Chronic granulomatous disease	3
22	Chediak-Higashi syndrome	2
23	Genetic agranulocytosis (i) Kostmann (autosomal recessive)	2
24	Genetic agranulocytosis (ii) autosomal dominant	1
25	Severe lazy leucocyte syndrome	1
26	Cyclic neutropenia	1
27	Adhesive protein deficiency	2
	Within phagocytes	
28	Osteopetrosis	6
29	Hurler's disease[2] (MPS IH)	10
30	Hunter's disease[2] (MPS II)	2
31	San Filippo A disease[2] (MPS IIIA)	1
32	San Filippo B disease[2] (MPS IIIB)	2
33	Morquio's disease[3] (MPS IVB)	2
34	Maroteaux-Lamy disease[3] (MPS VI)	1
35	Gaucher's disease type I	1

Table II. (cont.)

		n[1]
36	Gaucher's disease type III	6
37	Gaucher's disease type III (Norbottnian)	1
38	Fabry's disease	2
39	Pompé's disease	0
40	Wolman's disease	0
	Within red cells	
41	Fanconi's aplasia	12
42	Diamond-Blackfan syndrome	2
43	β-Thalassaemia major	12
44	Sickle cell anaemia	1
45	Hereditary spherocytosis	0
		212

[1] Estimated number of successful surviving patients.
[2] Not yet known if brain can recover.
[3] Not yet known if joint deformities achieve good normal function.

DBMT, the choice of donor, the displacement induction, GVHD and general care during DBMT is given elsewhere [6] but this short review considers the importance of immunoprophylaxis, the results obtained in specific lysosomal disorders and future possibilities.

Immunoprophylaxis

The human fetus becomes immunologically competent soon after 12 weeks of gestation. Apart from in vitro fertilisation techniques, current prenatal diagnosis programmes do not usually achieve a diagnosis before 12 weeks, and this means that intrauterine transplants or genetic engineering are faced with possible induction of an immune reaction against a new gene product which the fetus may not have had at the time that tolerance was being established. A fetus or an infant can make antibodies that can interfere with enzymes in two major ways:

Table III. Genetic causes of impaired enzyme

	Product	Antigenicity of normal enzyme
DNA sequence error		
Starter	nul	+
Out-of-step	nonsense	+
Splicing	nonsense	+
Translation	nonsense	+
Active site	variable	0 to +
Cho side-chain site	not in lysosome	0 to +
Related		
PO_4 or cho enzyme	not in lysosome	0
Coenzyme		0

(i) Most commonly, they bind to the enzyme, impairing its distribution and accelerating its catabolism.

(ii) Less frequently they actually block the active site of the enzyme, and we have observed this in 7 patients [6].

Antibodies of type (i) are more difficult to study, preferably requiring a purified enzyme whose half-life and distribution could then be followed.

The potential antigenicity of a donor component given to an affected patient is only partially predictable. Impairment of the enzyme activity of the recipient can arise in a number of different ways (see table III) and this can affect whether or not the recipient will be tolerant to the normal component. Clearly, where there is no enzyme protein at all (null gene) or a large change in the coded and, therefore, amino acid sequence, the normal enzyme will be immunologically much different from the patient's experience. Where such an immunocompetent patient receives any form of enzyme therapy, using red cells (e.g. adenosinedeaminase), leukocytes (even irradiated), fibroblasts, amniotic membrane cells, other organ grafts, or purified enzymes which may contain denatured molecules, there is a great risk that antibody responses will be engendered. Once these have clonally expanded beyond the primary divisions, it will be very difficult indeed to prevent them recurring following BMT. It is well known in clinical immunology that secondary antibody responses are very difficult to delete. On the other hand, primary antibody responses can be abrogated by the use of cyclophosphamide at 20 mg/kg body weight on day +1 and day +2 after first introducing the new component to the child. It is very important that this should be at least 24 h after the new protein has been given to the child, for this allows helper T cells to initiate the B cell proliferation and it is on the next 2 days that the B lymphocytes will be very susceptible to the cytoreductive treatment. It is therefore considered that such immunoprophylaxis is mandatory wherever a child who might require a DBMT has to be given any of the above products for the first time. During the preparation for the transplant we also carefully time the introduction of the donor buffy coat (containing the normal component) exactly 24 h before the cyclophosphamide course of the induction. Failure to use such immunoprophylaxis could account for the lack of success of several transplants reviewed in Hobbs [6], and in 4 of our earlier patients we have detected that type (ii) antibodies were, in fact, formed. When a DBMT is undertaken with the above immunoprophylaxis, it can produce 100% engraftment of the donor bone marrow cells which, in due course, will become the immunological repertoire of the recipient. The use of DBMT confers upon the patient the donor's tolerance of his own enzyme. This is almost a unique advantage among the many recent attempts at enzyme therapy.

Results in Lysosomal Disorders

The first successful transplant intended to replace the abnormal phagocytes of a patient by normal phagocytes, was done in April 1973 for chronic granulomatous disease [3]. At the time, it was thought that the deficiency was of a lysosomal enzyme, but it is now considered to be of a co-enzyme [7]. The nitroblue tetrazolium test identified the donor phagocytes which only reached 12% in the recipient and were the first clear indication for the need for displacement in such attempts. Subsequent grafts, using displacement manoeuvres, have corrected the phagocyte defect in this disease [4].

Chediak-Higashi Syndrome

Animal experimenters traditionally use total body irradiation, so by serendipity displaced the abnormal phagocyte lines, and cured a form of Chediak-Higashi syndrome in mice [8] and the disease has since been cured in man [9].

Osteopetrosis

This autosomal recessive disease is due to a failure of bone resorption [10] and bony overgrowth crushes cranial nerves to cause blindness, deafness, and other palsies and displaces the bone marrow. Allotypic variation seems to result in different times of onset and rate of progress, but death most commonly occurs in early childhood with only the occasional survival to 20 years. The osteoclast arises from monocyte precursors of the bone marrow [11] so that the elegant parabiotic studies of Walker [12] were superseded by BMT [13] to cure micro-ophthalmic mice. While in those animals DBMT did not seem to be necessary [14], initial attempts at infusing compatible cells [15] and BMT without displacement [16] failed in man; it is quite clear that, in man, no success occurs unless DBMT is used. In 1979, Lamendin [17] achieved bone resorption after using irradiation in the induction, later confirmed in America [18]. The recommended [4] use of Busulphan to avoid irradiation of infants has since been successful [19, 20] with careful control of the hypercalcaemia that follows the mobilization of the bone. As the child grows older, even DBMT can fail and it is not known whether this is due to the many previous blood transfusions (without immunoprophylaxis) or whether the donor marrow has been cleaned too thoroughly of donor monocytes which might be needed to establish the incoming stem cells in the bone spaces. For best results, if the disease is detected at 36 weeks of gestation, premature delivery might save the cranial nerves (avoiding blindness and deafness), aseptic isolation would reduce the risks of infection, and calcitriol could delay petrosis while a donor is arranged.

For the above diseases, while the defect clearly relates to the lysosome, it could be argued that their correction has occurred because the phagocyte/monocyte line has been replaced, and that enzyme transfer itself has not been necessary. Such an argument does not apply to the following diseases.

Gaucher's Syndrome

This includes infantile acute (type II) usually fatal within 1 year, juvenile subacute (type III) which is more variable and with neurological involvement aggravated after splenectomy [22], and the adult non-neurological (type I) where crippling deformities of the long bones precede final bone marrow failure. Enzyme replacement therapy in types I and III has not been impressive [23] but was conducted without immunoprophylaxis. Early BMT (using Busulphan) has been successful [23–25]. Irradiation achieved donor leucocytes containing beta-glucocerebrosidase [24] but immunoprophylaxis was unsatisfactory [6] and plasma correction was very slow; perhaps when the enzyme left the white blood cells, it met antibodies. Avoiding irradiation and using immunoprophylaxis [26] has now resulted in four successful grafts at Westminster; however, one patient has died of infection. Septicaemia is a serious problem in these patients as many have had splenectomy and it is now known that the spleen plays an important role in programming a subset of B lymphocytes [27] which probably later populate the bone marrow to endow it with memory. We have, therefore, adopted the practice of immunizing the prospective donor against pneumococci (one of the common offending organisms causing death) and then transferring this memory in the bone marrow transplant. It remains to be seen if this will reduce the incidence of post-splenectomy infection. Ideally, early diagnosis permitting, the DBMT should be undertaken before splenectomy becomes indicated, before diaphyseal bone growth has ceased (so that new bone unaffected by the previous enzyme deficiency can be formed) and before the bone marrow space is totally occupied by Gaucher deposits. In those patients who had pre-existing bony deformities, it does appear that the healing is slow, but surprisingly, the strengthening is remarkable after the graft. It is not known whether the type II disease can respond to DBMT and the reservations with regard to neurological involvement must be the same as those below for some of the mucopolysaccharidoses.

Fabry's Disease

While kidney transplants temporarily solve the renal problem for patients, they do not significantly improve the systemic disease elsewhere in the body. Fetal liver transplantation has had systemic beneficial effects in 3 patients [28] indicating that only a small amount of the normal α-galactosidase has value for this disease. Fetal liver contains the forerunner of the bone marrow and the advantage of DBMT is that it would give a much larger graft together with tolerance of the enzyme, and should last for the entire lifetime of the patient.

Wolman's Disease

Diagnosed early by Dr. C.H. Cheetham, a boy had the severe form of acid esterase deficiency with a previous infant having died at 14 weeks. A compatible sister who was normal was detected, but the parents hesitated until at 10 weeks the boy was deteriorating with diarrhoea, vomiting and massive hepatosplenomegaly. Peripheral leucocytes, liver and bone marrow biopsies showed gross cholesteryl ester inclusions. DBMT achieved 100% female neutrophils with a normal enzyme content, and by day +21 large amounts of cholesterol had been mobilized; the plasma level rose from 3.23 to 11 m*M*/l to slowly fall to 5.9 m*M*/l by day +68.

Liver and bone marrow biopsies showed a reduction of the inclusions, and his vomiting and diarrhoea ceased. Serum bilirubin peaked at 610 m*M*/l as it had been a blood group 'A' into 'O' graft, but this had normalized by day +68 and accompanying ascites had also diminished. He gained weight and appeared to be thriving, with a good appetite, a smaller and softer liver and spleen, but suddenly deteriorated on day +76. *Staphylococcus epidermidis* was isolated from his blood culture and despite treatment for this, he died on day +80. The post-mortem revealed widespread aspergillosis. DBMT, while correcting his genetic disease, had alas predisposed to his fatal infection.

The Mucopolysaccharidoses (MPS)

Further investigation of infants presenting with hernia can early detect this group of diseases [29], hopefully before brain damage becomes irreversible. Early treatment could allow the growth of brain connections in the post-graft period, believed to continue to at least the age of 3 years. From the evidence below, the mental deterioration can be halted, but only long-term follow-up will indicate if the eventual mental state justifies the risks of the procedures in any of these terrible diseases. At Westminster Children's Hospital, 5 years of follow-up are encouraging us to continue. Certainly most of the systemic effects of these diseases can be corrected, the DBMT done under elective conditions (a 'clean' recipient), and the transplant survival is better than for many leukaemia and aplasia patients [6]. Nevertheless, whether the IQ of these children will eventually be adequate for them to lead a normal life, remains an open question.

This will be important in finally deciding, in the absence of any satisfactory alternative, whether such treatment is continued for such conditions affecting the brain. Some enzyme does enter the brain; it seems unlikely that it crosses the blood-brain barrier as free protein, but more probable that it is taken across by diapedesing cells; whether lymphocytes or monocytes [30] are more important is unknown. New techniques of opening the blood-brain barrier are being developed and explored and could increase the delivery of enzyme where there has been successful engraftment.

Hurler's Disease (MPS I)

Of 16 patients treated by DBMT, 8 survived, 1 regrafted after 1 year. Three children transplanted from donors with normal enzyme levels have all shown rapid normalization of liver, spleen and corneal clouding within 3 months. By 18 months they could flatten their fingers together in the 'prayer' position. While they stood up straight and appeared to develop normally there has been no significant radiological change in the 'beaked' vertebra at the main point of the original lumbar gibbus, but its angle has not increased as in untreated children. Subluxation of the hips of one child spontaneously resolved within 9 months postgraft, so that a planned operation could be cancelled. The accelerated skull growth ceased in all [31], possibly due to rapid relief of hydrocephalus in some cases, but there has also been improvement in both CAT scans and in NMR scans [32] with considerable differentiation between myelinated white and grey matter (not seen in a few untreated MPS patients). In one boy who died 15 months postgraft from pneumococcal septicaemia, the brain substance contained α-iduronidase at 4% of the level in a normal brain, whereas untreated MPS I brain contains no enzyme activity. His brain did not show complete clearing of MPS whereas the liver did, as previously reported on biopsy [33]. Another pleasing result has been the slow normalization of facial appearance, and in two boys now 4.5 and 5 years postgraft, the original diagnostic facies is no longer obvious.

Where the donor has been a presumed heterozygote with half the enzyme level, the impression has been of slower clearing of the eyes (4–6 months) and even at 3 years the fingers cannot be voluntarily straightened. In summary, 7 survivors enjoy good health, their parents are delighted with their progress, the systemic disease is certainly controlled by the transplant, but the all important question of their eventual mental status remains to be evaluated. Four patients are now attending normal schools and appear to be keeping up with the other children.

Hunter's Disease (MPS II)

Three patients survive from the severe form of the disease, two have problems; one has chronic GVHD and is on azathioprine with low levels of leucocytes and enzyme; the other is a mixed chimaera of unknown stability (30% male leucocytes, 70% female donor) and enzyme delivery may be suffering. Some time will be needed to assess their progress which, to date has not been impressive. Allotypic variation can be found with graded rates of progress of the disease from very fast to very slow [34]. Since dendritic growth within the brain is now believed to have largely ceased by the age of 3 years, the eventual mental status must be related to whether or not the transplanted patients had that benefit (grafted before the age of 2 years) or whether, indeed, they were treated at a later age with little possibility of change. Since there is no experimental model in animals, evidence can only be collected by undertaking treatment of more Hunter children, preferably under 2 years, and from matched siblings with normal enzyme levels.

Sanfilippo Diseases (MPS III A, B, C)

This group of diseases is probably among the most unpleasant to afflict family life, children becoming hyperactive, destructive and unmanageable. The first type B family we encountered was so moving that attempting DBMT was justifiable [34]. The progress of the non-identical twins, transplanted at the same time, was complicated by severe GVHD, with leucopenia and low enzyme levels, yet it has been much better than that of their two older, untreated, brothers. Four years post-graft the twins are 6 years old, with correction of hepatosplenomegaly and though somewhat hyperactive, are learning new skills. Thus, it seems justified to continue to try and help such children.

It is too early to assess the progress of the Sanfilippo 'A' boy who survived his transplant, other than to say that the systemic features of the disease have improved but he has not shown any mental improvement.

Morquio's Disease (MPS IV)

Fortunately, in this group, where again, allotypic variation is now realized to exist [34], the brain is usually unaffected. Because the bone derangements become so serious and create bad muscular habits, it would be preferable to treat these children as soon as a definitive diagnosis has been made. While the liver, spleen and eyes improve [36], there is so far little improvement in the joint abnormalities. The first patient, (a girl) has produced an IgG that binds Morquio enzyme, partially lowering its activity and probably impairing its distribution. The second patient (a boy) has had chronic GVHD with marked leucopenia after a graft from his father (heterozygote), so enzyme levels have unfortunately been low.

Maroteaux-Lamy Syndrome (MPS VI)

A patient has been transplanted for this disease [37], obtaining improvement in hepatosplenomegaly and the cornea; the progress of the joint deformities has the same provisos as for MPS IV.

Pompé's Disease

In this disease, a deficiency of acid maltase results in the accumulation of glycogen, especially within the muscles of the body, impairing their function. Gross enlargement of the liver, spleen and heart occur, and death is most often due to cardiac failure. Infusing preparations of acid maltase has reduced the amount of glycogen in the liver, spleen and voluntary muscle. In these organs coated pits and endosomes are readily detectable, permitting entry of donor enzyme, and certainly endosomes are visible within cardiac muscle, but it is not known if the coated pits are lost after birth. An infant survived for 98 days after DBMT which established circulating leucocytes with donor enzyme [to be published] but after death from infection and heart failure, it was considered that no glycogen had been mobilized from cardiac muscle. In that graft, however, immunoprophylaxis was not practised, and no glycogen was removed from liver, spleen or voluntary muscle. It remains to be seen whether immunoprophylaxis will permit delivery of the enzyme.

Results and Discussion

Of 70 infants and children, with otherwise fatal genetic disease, who have received transplants at Westminster, 30 have been restored to leading normal lives, with our longest survivor having done so for 15 years. A further 7 children are alive, but 3 are affected by chronic GVHD and 4 are still suffering from the residual effects of the advanced state of their disease at the time of their transplant. A further 10 children have died from lesions incurred before the transplant was undertaken, e.g. bronchiectasis, and would not really qualify as elective transplants. Deaths from acute GVHD have numbered 8, and for 7 of those, the graft was known to be mismatched, but the risk was taken in the hope of being able to combat the GVHD. A further 15 patients have died of infection, most often within 1 year of the transplant before the immunity postgraft has become fully normalized; in 4 of these 15 patients, the infection was associated with chronic GVHD. There are, therefore, 23 deaths which can be directly attributed to undertaking the transplant procedure, though in 18 of these it was a graft from a mismatched or unrelated donor where there is a high risk.

Where DBMT is undertaken under elective conditions, and before the disease has become too advanced, the current success rate can be expected to be 80–90% [6]. Where the donor is a half-matched family member other

Table IV. Inborn errors of metabolism possibly correctable by bone marrow transplantation

	Of lymphocyte function
46	Combined immunodeficiency in adults <40 years
47	Bruton's hypogammaglobulinaemia
48	Sporadic hypogammaglobulinaemia
49	Isolated IgM deficiency
50	Genetic isolated IgG deficiency
51	Severe IgG subclass deficiency
52	Severe IgA deficiency
53	X-linked proliferative syndrome
54	Orotic aciduria
	Of phagocytosis
55	Severe myeloperoxidase deficiency
56	Homozygous G6PD deficiency
57	Lipochrome histiocytosis
58	Lysozyme deficiency
59	Lactoferrin deficiency
60	Strauss-type defect
61	Defect in tubulin
62	Defect in myosin
63	Severe histiocytosis-X
64	Familial reticuloendotheliosis (D. Miller)
	Within phagocytes
65	Scheie's disease
66	Hurler-Scheie compound (MPS ISH)
67	San Filippo C (MPS IVc)
68	San Filippo D (MPS IVd)
69	β-Glucuronidase deficiency (MPS VII)
70	Metachromatic leucodystrophy
71	Adrenoleucodystrophy
72	GM1-gangliosidosis
73	Mucolipidosis II
74	Mucolipidosis III
75	Farber's lipogranulomatosis
76	Niemann-Pick A disease
77	Niemann-Pick C disease
78	Niemann-Pick D disease
79	Mannosidosis
80	Fucosidosis
81	Sialidosis
82	Krabbe's disease
83	Urea cycle deficiencies

Table IV. (cont.)

84	Maple-syrup urine disease
85	Galactosaemia
86	Refsum's disease
87	Gaucher's Type II disease
88	Lesch-Nyhan syndrome
89	α_1-Antitrypsin deficiency
90	Haemophilia
91	Ataxia telangiectasia
92	Xeroderma pigmentosa
	Within red cells
93	Severe enzyme deficiency
94	Other severe haemoglobinopathies
95	Haemopoietic porphyria

than the father or mother, with an elective graft undertaken in a unit fully equipped with laminar flow facilities, and using modern methods of anti-GVHD, the success rate can now approach 60% [6].

Future Possibilities

In table IV are listed a further 50 diseases where it is suggested that if a fully compatible family donor with a normal level of the missing component is available, DBMT could be considered. It has already been attempted in metachromatic leucodystrophy [38] and in a surviving patient [39] DBMT has achieved circulating donor leucocytes, but after failed previous attempts without immunoprophylaxis; some success has been achieved and it will be interesting to see if, in the future, DBMT with immunoprophylaxis can achieve better results. The same can be said for adrenoleucodystrophy. In GM1-gangliosidosis, however, there have now been two attempts at DBMT, without any significant improvement [6].

Tissue culture of the prospective recipient's biopsies does enable in vitro studies to be made, adding the prospective donor's leucocytes, to see whether correction can be achieved in vitro and elegant studies of this type have been done [40–43], setting a pattern for the future. Such a study avoids the possibility that a purified enzyme may have been damaged during its preparation; naturally occurring Hunter-corrective factor (from the urine) does correct Hunter fibroblasts, whereas enzyme extracted from the plasma has failed to do so [44]. Improvements in the liver and spleen in our own treated patients suggest that in vivo transfer does occur, but the all-important question of whether enough gets in to the brain cells still needs an answer. It may be possible to keep some nervous tissue (e.g. in a rectal biopsy) alive long enough to conduct a suitable in vitro study.

It is also very important, especially in small experimental animals, that those who undertake transplant experiments practise immunoprophylaxis: this might resolve some of the already conflicting findings between man and animals.

Mention should also be made of the potential cost-benefit that can ensue from DBMT. We have estimated that in England and Wales there are some 350 infants born each year who could be treated with DBMT; who at present cost our Health Service nearly £6.5 million p.a. and whose transplants would cost a maximum of £2 million p.a. and undoubtedly would relieve a very great deal of suffering that occurs, not only for the patients but also for their families.

Acknowledgements

The work of the Westminster Bone Marrow Transplant Team has also involved a large number of other colleagues listed in Hobbs [5]. We are grateful for generous financial help from the Bostic, Fane and Dobson Funds of the Westminster Research Trust, the Cancer Research Campaign and the 'Woman's Own' Save a Child Fund.

References

1 Yamamura, M.; Newton, R.; James, D.; Humble, J.; Butler, L.; Hobbs, J.: Uncomplicated HL-A matched sibling bone marrow graft for combined immune deficiency. Br. med. J. *ii:* 265–269 (1972).

2 Hobbs, J.; Humble, J.; Anderson, I.; James, D.: The elective treatment of graft-versus-host disease following a bone marrow graft from a father to a son with severe combined immunodeficiency. Post-grad. med. J. *52:* suppl. 5, pp. 91–95 (1976).

3 Foroozanfar, N.; Hobbs, J.; Hugh-Jones, K.; Humble, J.; James, D.; Selwyn, S.; Watson, J.; Yamamura, M.: Bone marrow transplantation from an unrelated donor for chronic granulomatous disease. Lancet *i:* 210–213 (1977).

4 Hobbs, J.: Bone marrow transplantation for inborn errors. Lancet *ii:* 735–739 (1981).

5 Hobbs, J.: The scope of allogeneic bone marrow transplantation; in Losowsky, Bolton, Advanced medicine Leeds, pp. 378–391 (Pitman, Bath 1983).
6 Hobbs, J.: Correction of 34 genetic disease by displacement bone marrow transplantation. Plasma Ther. Transfus. Technol. *6:* 221–246 (1985).
7 Segal, A.; Jones, O.: Absence of cytochrome b reduction in stimulated neutrophils from both female and male patients with chronic granulomatous disease. FEBS Lett. *110:* 111–114 (1980).
8 Kazmierowski, J.; Elin, R.; Reynolds, H.; Durbin, W.; Wolff, S.: Chediak-Higashi syndrome: Reversal of increased susceptibility to infection by bone marrow transplantation. Blood *47:* 555–559 (1976).
9 Virelizier, J.; Durandy, A.; Lagrue, A.; Fischer, A.; Griscelli, C.: Successful bone marrow transplantation in a patient with Chediak-Higashi syndrome. Exp. Hematol., Copenh. *11:* suppl. 13, pp. 91–92 (1983).
10 Fraser, D.; Kooh, S.; Chan, A.; Cherian, A.: Congenital osteopetrosis. A failure of normal resorptive mechanisms of bone. Calcif. Tissue Res. *2:* suppl., p. 526 (1968).
11 Teitelbaum, S.; Kahn, A.: Mononuclear phagocytes, osteoclasts and bone resorption. Mineral Electrolyte Metab. *3:* 2–9 (1980).
12 Walker, D.: Bone resorption restored in osteopetrotic mice by transplants of normal bone marrow and spleen cells. Science *190:* 784–785 (1975).
13 Loutit, J.; Sansom, J.: Osteopetrosis of microphthalmic mice. A defect of the haemopoietic stem cell? Calcif. Tissue Res. *20:* 251–259 (1976).
14 Loutit, J.: Bone marrow grafts in mature osteopetrotic mice. Transplantation *24:* 299–301 (1977).
15 Ballet, J.; Griscelli, C.; Coutris, C.; Milhaud, G.; Maroteaux, P.: Bone marrow transplantation in osteopetrosis. Lancet *ii:* 1137 (1977).
16 Sorrell, M.; Rosen, J.; Kapoor, N.: Bone marrow transplant for osteopetrosis in a 10-year-old boy. Pediat. Res. *13:* 481 (1978).
17 Lamendin, H.: First case of bone resorption evoked by bone marrow transplantation in a human patient with osteopetrosis. Mater. med. pol. *11:* 67–68 (1979).
18 Coccia, P.; Krivit, W.; Cervenka, J.; Clawson, C.; Kersey, J.; Kim, T.; Nesbit, M.; Ramsay, N.; Warkentin, P.; Teitelbaum, S.; Kahn, A.; Brown, D.: Successful bone marrow transplantation for infantile malignant osteopetrosis. New Engl. J. Med. *302:* 701–708 (1980).
19 Sorrell, M.; Kapoor, N.; Kirkpatrick, D.; Rosen, J.; Chaganti, R.; Lopez, C.; Dupont, B.; Pollack, M.; Terrin, B.; Harris, M.; Vine, D.; Rose, J.; Goossen, C.; Lane, J.; Good, R.; O'Reilly, R.: Marrow transplantation for juvenile osteopetrosis. Am. J. Med. *70:* 1280–1287 (1981).
20 Sieff, C.; Levinsky, R.; Rogers, D.; Muller, K.; Chessells, J.; Pritchard, J.; Casey, A.; Hall, C.: Allogeneic bone marrow transplantation in infantile malignant osteopetrosis. Lancet *i:* 437–441 (1983).
21 Key, L.; Carnes, D.; Cole, S.; Holtrop, M.; Bar-Shavit, Z.; Shapiro, F.; Arceci, R.; Steimberg, J.; Gundberg, C.; Kahn, A.; Teitelbaum, S.; Anast, C.: Treatment of congenital osteopetrosis with high-dose Calcitriol. New Engl. J. Med. *310:* 409–415 (1984).
22 Svennerholm, L.; Mansson, J.E.; Nilsson, O.; Tibblin, E.: Bone marrow transplantation in the Norrbottnian form of Gaucher disease; in Barranger, Brady, Molecular

basis of lysosomal storage disorders, pp. 441–459 (Academic Press, New York 1984).
23 Brady, R.: Enzyme replacement in the sphingolipidoses; in Barranger, Brady, Molecular basis of lysosomal storage disorders, pp. 461–478 (Academic Press, New York 1984).
24 Ginns, E.; Caplan, D.; Rappeport, J.; Barranger, J.: Bone marrow transplantation in severe Gaucher's disease: rapid correction of enzyme deficiency accompanied by continued long term survival of storage cells; in Barranger, Brady, Molecular basis of lysosomal storage disorders, pp. 429–440 (Academic Press, New York 1984).
25 Hobbs, J.; Hugh-Jones, K.; Shaw, P.; Poynton, C.: Displacement bone marrow transplantation in adult-type Gaucher's disease (Abstract). Br. paediat. Ass. Meet., 1984.
26 Hobbs, J.: Bone marrow transplantation for some inborn errors; in Crawfurd, Gibbs, Watts, Advances in the treatment of inborn errors of metabolism, pp. 109–121 (Wiley, Chichester 1982).
27 Waldschmidt, T.; Layton, J.; McFadden, S.; Uhr, J.; Vitetta, E.: $slgD^-$ and $slgD^+$ B cells responding to thymus-independent antigens have a different lymphokine requirement. J. Immun. *135:* 865–867 (1985).
28 Touraine, J.L.: Fetal liver transplantation in congenital enzyme deficiencies in man. Exp. Hematol., Copenh. *10:* suppl. 10, p. 46 (1982).
29 Hugh-Jones, K.: Early diagnoses in mucopolysaccharidoses. Lancet *ii:* 1300 (1983).
30 Oehmichen, M.: Are resting and/or reactive microglia macrophages? Immunology *161:* 246–254 (1982).
31 Hugh-Jones, K.; Hobbs, J.; Chambers, D.; White, S.; Byrom, N.; Williamson, S.; Barrett, J.; Henry, K.: Bone marrow transplantation in mucopolysaccharidoses; in Barranger, Brady, Molecular basis of lysosomal storage disorders, pp. 411–428 (Academic Press, New York 1984).
32 Johnson, M.; Desai, S.; Hugh-Jones, K.; Starer, F.: NMR Imaging of the brain in Hurler's Syndrome. Am. J. Neuroradiol. (in press, 1986).
33 Hobbs, J.; Barrett, A.; Chambers, D.; James, D.; Hugh-Jones, K.; Byrom, N.; Henry, K.; Lucas, C.; Roigers, T.; Benson, P.; Tansley, L.; Patrick, A.; Mossman, J.; Young, E.: Reversal of clinical features of Hurler's disease and biochemical improvement after treatment by bone marrow transplantation. Lancet *ii:* 709–712 (1981).
34 Orii, T.: Ministry of Education report on 850 patients with mucopolysaccharidoses 1979–1984 (Gifu University, Japan 1984).
35 Hugh-Jones, K.; Kendra, J.; James, D.; Rogers, T.; Williamson, S.; Desai, S.; Patrick, A.; Hobbs, J.: Treatment of Sanfilippo B disease K(MPS IIIB) by bone marrow transplant. Exp. Hematol., Copenh. *10:* suppl. 10, pp. 50–51 (1982).
36 Desai, S.; Hobbs, J.; Hugh-Jones, K.; Williamson, S.; Barnes, I.; Kendra, J.; White, S.: Morquio's disease (mucopolysaccharidosis type IV) treated by bone marrow transplant. Exp. Hematol., Copenh. *11:* suppl. 13, pp. 98–100 (1983).
37 Krivit, W.: Correction of inborn errors of metabolism by bone marrow transplant; in Hollaender, Laskin, Rogers, Basic biology of new developments in biotechnology, pp. 63–76 (Plenum Press, New York 1983).
38 Joss, V.; Rogers, T.; Hugh-Jones, K.; Beilby, B.; Joshi, R.; Williamson, S.; Foroozanfar, N.; Riches, P.; Turner, M.; Benson, P.; Hobbs, J.: A bone marrow transplant for metachromatic leucodystrophy. Exp. Hematol., Copenh. *10:* suppl. 10, pp. 52–55 (1982).

39 Bayever, E.; Philippart, M.; Nuwer, M.; Ladiaxh, S.; Brill, N.; Sparkes, R.; Feig, S.: Bone-marrow transplantation for metachromatic leucodystrophy. Lancet *ii:* 471–473 (1985).
40 Wiseman, U.; Rossi, E.; Herschkowiz, N.: Treatment of metachromatic leukodystrophy in fibroblasts by enzyme replacement. New Engl. J. Med. *284:* 672 (1971).
41 Porter, M.; Fluharty, A.; Kihara, H.: Correction of abnormal cerebroside suflate metabolism in cultured metachromatic leukodystrophy fibroblasts. Science *172:* 1263–1265 (1971).
42 Neufeld, E.; Cantz, M.: In Hers, Van Hoof, Lysosomes and storage disease, p. 261 (Academic Press, New York 1973).
43 Oldon, I.; Dean, K.; Harris, G.; Muir, H.: Direct transfer of a lysosomal enzyme from lymphoid cells to deficient fibroblasts. Nature, Lond. *291:* 244–247 (1981).
44 Wasteson, A.; Neufeld, E.: Iduronate sulfatase from human plasma. Meth. Enzymol. *83:* 573–578 (1982).

J.R. Hobbs, MD, Chairman, Westminster Bone Marrow Team, Department of Chemical Immunology, Charing Cross and Westminster Medical School, Westminster Hospital, London SW1P 2AR (UK)

Subject Index